# Springer Theses

Recognizing Outstanding Ph.D. Research

## Aims and Scope

The series "Springer Theses" brings together a selection of the very best Ph.D. theses from around the world and across the physical sciences. Nominated and endorsed by two recognized specialists, each published volume has been selected for its scientific excellence and the high impact of its contents for the pertinent field of research. For greater accessibility to non-specialists, the published versions include an extended introduction, as well as a foreword by the student's supervisor explaining the special relevance of the work for the field. As a whole, the series will provide a valuable resource both for newcomers to the research fields described, and for other scientists seeking detailed background information on special questions. Finally, it provides an accredited documentation of the valuable contributions made by today's younger generation of scientists.

## Theses are accepted into the series by invited nomination only and must fulfill all of the following criteria

- They must be written in good English.
- The topic should fall within the confines of Chemistry, Physics, Earth Sciences, Engineering and related interdisciplinary fields such as Materials, Nanoscience, Chemical Engineering, Complex Systems and Biophysics.
- The work reported in the thesis must represent a significant scientific advance.
- If the thesis includes previously published material, permission to reproduce this must be gained from the respective copyright holder.
- They must have been examined and passed during the 12 months prior to nomination.
- Each thesis should include a foreword by the supervisor outlining the significance of its content.
- The theses should have a clearly defined structure including an introduction accessible to scientists not expert in that particular field.

More information about this series at http://www.springer.com/series/8790

William M.R. Simpson

# Surprises in Theoretical Casimir Physics

## Quantum Forces in Inhomogeneous Media

Doctoral Thesis accepted by
the University of St Andrews, Scotland

Springer

*Author*
Dr. William M.R. Simpson
Department of Physics of Complex Systems
The Weizmann Institute of Science
Rehovot
Israel

*Supervisor*
Prof. Ulf Leonhardt
Department of Physics of Complex Systems
The Weizmann Institute of Science
Rehovot
Israel

ISSN 2190-5053 ISSN 2190-5061 (electronic)
ISBN 978-3-319-36400-1 ISBN 978-3-319-09315-4 (eBook)
DOI 10.1007/978-3-319-09315-4

Springer Cham Heidelberg New York Dordrecht London


Softcover reprint of the hardcover 1st edition 2015


Printed on acid-free paper

Springer is part of Springer Science+Business Media (www.springer.com)

וַיֹּאמֶר אֱלֹהִים יְהִי־אוֹר וַיְהִי־אוֹר׃

—Bereshit, Torah

*And God said: 'Let there be light.'*
*And there was light.*

—Genesis, Torah

*For my father,*
*Anthony Peter Simpson*
*1948–2009,*
*in a season of 'bright sadness'.*

# Supervisor's Foreword

At the beginning of the twentieth century, Albert Einstein replaced the aether theory with relativity. However, a twenty-first century aether is still puzzling physicists today: it is called the quantum vacuum.

The quantum vacuum is the ground state of the physical fields. Although the average of the electromagnetic field is zero in its ground state, its amplitude still fluctuates. These vacuum fluctuations give rise to the 'Casimir force' between dielectric bodies—a sticky force that is ubiquitous throughout nature. It causes parking tickets to attach to windscreens and the moving parts of micromachinery to attract and stick together.

Despite theoretical work for more than half a century, this 'Casimir effect' is not as well understood as one might suppose. In this thesis, the author uncovers new puzzles and paradoxa concerning this mysterious phenomenon. In particular, he clearly demonstrates that the most sophisticated modern theories of the Casimir force fail when they are confronted with dielectrics where the refractive index is not uniform but changes smoothly.

This thesis opens new vistas for research, but it also expounds the known foundations of Casimir theory at the level of a good textbook. Students or interested researchers can easily follow the discussion and learn the state of the art, as *art* it is.

The author, William Simpson, was a Ph.D. graduate student at the University of St Andrews in the UK, and a visiting student at the Weizmann Institute of Science in Israel; I had the pleasure of being his supervisor. During his research, he also spent several months at The University of Trento in Italy where he learned from Prof. Lev Pitaevskii, one of the early pioneers of Casimir theory. William is now a postdoctoral fellow at The Weizmann Institute.

In addition to Physics, William has a wide range of interests and talents. He reads theology and philosophy; he writes computer programs and composes poetry. The philosopher Roger Scruton, recalling William's movements among the sciences and the humanities, described him as 'a serious and imaginative thinker… (and) a man capable of organising and leading others in the pursuit of truth'. His thesis is written in clear, beautiful English—a delight to read.

Rehovot, May 2014 Prof. Ulf Leonhardt

# Preface

Credo ut intelligam.[1]

—St Anselm of Canterbury, Proslogion, 1

This is a book on quantum forces. More specifically, it is about predicting certain kinds of forces that arise between macroscopic bodies, for which we require quantum electrodynamics. More precisely, it is about the problem of predicting the size and nature of these quantum-mechanical forces when they arise within inhomogeneous media, in which the optical properties of a material are continuously changing as a function of position. However, in attempting to apply existing theory to such cases—and the general presumption is that it ought to be applicable—we uncover a number of *surprises*. In a sense, the basic thrust of this thesis is not so much to persuade its readers of certain answers as to convince them of unresolved questions. Some precise solutions to new problems are to be found within these pages, each of them interesting in its own right, but they are representative of a rather small class of Casimir problems involving inhomogeneous media that can be solved at present.

I hope this provocative paragraph will prove sufficiently intriguing to induce the turning of a few more pages, as I mean to avoid disclosing too many 'plot spoilers'.[2] I should warn the reader that it is also my preference to introduce formal physical theory in much the same way as I have learned it when left to my own devices—that is, as and when it is needed. While certain things are assumed to be familiar (some basic electrodynamics and some quantum mechanics, for instance), I occasionally invoke ideas or results beyond the common core of a Bachelor degree in Physics that are not 'fleshed out' until a subsequent section, when I think I can get away with doing so. For the more axiomatically minded, this may cause some minor abrasion. Nevertheless, to insist on understanding everything before we will say anything *about* it, do anything

[1] Translated: I believe in order that I may understand.

[2] For a summary of the conclusions, see the Outlook.

*with* it, or invest some kind of belief *in* it, is a limit in which nothing may ever be said or done again; the heat-death of the universe is the state that requires the most information, but the prospects for doing Physics at this point seem rather bleak.

## Outline of Chapters

This thesis is divided up as follows. In Part I, the basic foundations of Casimir Physics are introduced to the reader, consisting of Casimir's groundbreaking thought-experiment, the prediction of the original Casimir force, a discussion of dispersion forces in general, select experimental verifications of Casimir phenomena, and the generalisation of the Casimir Effect to arbitrary geometries and more realistic materials using the basic framework of macroscopic quantum electrodynamics.

In Chap. 1, we consider Casimir's original thought-experiment, involving two perfect mirrors in a vacuum at zero temperature. Casimir's prediction of a finite, attractive force is recovered, following a discussion of how to regularise (and then renormalise) the infinite zero-point energy of the electromagnetic field, whose 'quantum fluctuations' are claimed to be the cause of it. The Casimir force, in fact, is part of a family of dispersion forces that Casimir studied, ubiquitous in nature, and we take a few moments here to survey the broader field, touching on both van der Waals and Casimir-Polder phenomena. The chapter is concluded by briefly reviewing some of the important experimental verifications of the Casimir effect.

Chapter 2 presents an outline of some of the basic elements of the theory of macroscopic quantum electrodynamics (macro-QED), affording a formal basis for ideas alluded to in Chap. 1, and deriving some of the principal results that will be used throughout this discussion, including the Lifshitz formula for the force between two separated dielectric half-spaces, and the ground-state energy of a quantum field. A new argument is offered for the correct form of the Casimir-Lifshitz stress tensor. The discussion, however, remains phenomenologically driven: the additional lengths we would be obliged to go to in presenting a properly *canonical* theory of macro-QED are not deemed to pay sufficient dividends for our restricted purposes, and may be studied elsewhere.

With the basics of macro-QED in place, Chap. 3 offers a brief excursus on the subject of the disputed *nature* of the Casimir force, which has been described, on the one hand, as an effect resulting from the alteration of the zero-point electromagnetic energy, by the imposition of external boundary conditions, and, on the other hand, as simply a giant van der Waals force between the metal plates. A different perspective is put forward, in which the Casimir force is seen to arise from the fluctuations of a polariton field involving the coupled, quantised system of dielectric material and electromagnetic fields. In its ground state, the system cannot be separated into material or electromagnetic quanta, which arguably splits the debate about the nature of the force straight down the middle.

In Parts II and III, we consider the case of the Casimir force in inhomogeneous media. Four peer-reviewed calculations are presented, resulting from this research, the first two focussed on the general problem of extending present theory to the case of macroscopic bodies in which the optical properties are varying continuously (the *surprises*), and the second two targeting exceptional cases of inhomogeneous media where the Casimir force can be determined exactly using ideas from transformation optics, leading to some interesting questions (the *conundrums*).

In the beginning, we considered the case of the Casimir force between two parallel plates separated by empty space, and subsequently saw that the force could also be calculated for the case in which the cavity is filled with a liquid medium. Chapter 4 asks the simple question of what happens to the force when the medium between the plates is *inhomogeneous*; that is, when the refractive index profile of the interposing liquid varies continuously as a function of position. A calculation is presented, based on Casimir's original approach to the cavity problem, but introducing the simple modification of a spatially dependent permittivity between the plates. A finite prediction for the Casimir force appears to be possible using a simple mode summation.

Chapter 5 reconsiders the same problem, only this time using the more sophisticated apparatus of Lifshitz theory, in which the detailed dispersive behaviour of the medium can be incorporated. The stress tensor inside the medium is determined using a piece-wise approximation, which is then taken to the continuum limit. However, the predicted force is now surprisingly illusive, and the possible need for incorporating additional information about the system, and perhaps more of the microphysical properties of the liquid, is discussed.

Noting the pathological nature of the Casimir-Lifshitz stress in an inhomogeneous medium, Chap. 6 considers the possibility of introducing an idealised inhomogeneous medium in the chamber that effectively modifies the size of an empty cavity. In this case, the Casimir force should be predictable and finite. The apparent contradiction with the previous chapter (Chap. 5) is explored and resolved, and in doing so we are able to determine an exact expression for the Casimir force for the case of a '*C*-slice' that (theoretically) reduces quantum stiction between attractive surfaces.[3]

In Chap. 7, we consider the case of the Casimir stress in Maxwell's fisheye—another inhomogeneous metamaterial with some remarkable properties. The stress tensor is infinite everywhere throughout the medium. However, a simple alternative regularisation is motivated, resulting in a finite stress tensor, leading to some perplexing questions about the nature of regularisation and our current understanding of the Casimir force.

Our journey ends with a survey of what we have learned in the *Outlook*, some suggestions for how to take this work further forward, and the proposal of a new experiment. I hope you enjoy the ride.

Rehovot, February 2014 Dr. William M.R. Simpson

[3] The phenomenon of quantum stiction leads to technological difficulties for micro and nano-electromechanical devices.

# Acknowledgments


*εν παντι ευχαριστειτε τουτο γαρ θελημα Θεου*[4]

—St Paul of Tarsus, 1 Thessalonians 5

Before beginning my acknowledgments, I should make some reasonable effort to restrict their domain; there are a great many creditable agencies in the world who have been causally relevant to the completion of this thesis, from the distant act of creation to the present moment of composition, and it would hardly be possible to refer to them all. I will confine myself, then, to mentioning those more immediately involved in producing the necessary conditions for a successful course of study. The quest for non-arbitrary 'cut-offs' in summing over large quantities, after all, represents a good part of my concern in the calculations that follow, and I might as well start as I mean to go on.

First, I would like to thank my supervisor, Prof. Ulf Leonhardt (The Weizmann Institute), for his extraordinary generosity during the course of my research, and for consistently treating me as a colleague instead of an underling. It is a dismal fact that Physics today is generally practiced in the ugliest building on the campus by bejeaned technicians starved of sunlight and maintained on coffee-making life-support machines. There can be few supervisors, I suspect, that take their students out to the opera or for concerts at the *Musikverein*. For many of the postgraduates that I have met on my travels, student life in the paper-factory world of modern academia seems to have degenerated into a kind of serfdom. I am grateful for the degrees of freedom and remarkable privileges I have been accorded by contrast.

Secondly, I am indebted to Dr. Simon Horsley (The University of Exeter), for his friendship, collaboration and borderline-obsessive interest in anything to do with Physics. Without his punctuated periods of intense involvement, and our many discussions over the course of the last year, Chap. 4 would not exist, and Chap. 5


---

4 Translated: In all [circumstances] give thanks, for this is the will of God.

might be less convincing. Coming out of a background in Solar Physics, with a skill-set lopsided towards numerics rather than analytics, I had a lot to learn as well as put into practice for the first time; our interactions helped me to start learning and doing, in the only way (I suspect) that we ever learn anything worthwhile or ever have done—'the whole theory of modern education is radically unsound'.

Thirdly, I should acknowledge my intellectual debts to a number of Physicists in the fields of Casimir Physics and beyond: Prof. Lev Pitaevskii (The University of Trento), of 'Landau and Lifshitz' fame, who kindly supervised a period of study in Italy, and who is doubtless still sadly waiting for me to produce 'the pressure term'; Dr. Thomas Philbin (The University of Exeter), who has helpfully answered my questions on numerous occasions, but hopes in vain that scientific materialism will some day answer the rest of them; Prof. Kenyon Chen (Soochow University), who most graciously hosted my visit to Suzhou in China, and pressed me to undertake the generalisation I present in Chap. 4; Mr. Itay Griniasty (The Weizmann Institute), for his repeated efforts to undermine my conclusions and his skiing lessons in Les Houches, Dr Ephraim Shahmoon (The Weizmann Institute), for helpful conversations concerning van der Waals phenomena, and Mr. Fanglin Bao (Zhejiang University) for finding a significant mistake that enabled me to find a significant result. I would also like to thank Dr. Scott Robertson (Université Paris-Sud), Mr. János Perczel (MIT) and Dr. Natalia Korolkova (The University of St Andrews) for their incisive comments on several chapters of this volume.

Fourthly, reaching beyond the world of Physics research, I am grateful for the friendship of Prof. Hans Halvorson (Princeton) and Prof. Raymond Tallis (a public intellectual), who continue to support my forays into the philosophy of Physics, and have enthusiastically participated in my 'PhysPhil' conferences, which have provided one source of motivation for my continued efforts in Theoretical Physics. This also leads me to thank fellow 'PhysPhils' Dr. Paul Rimmer, Ms. Janice Chik, Dr. Stephen Gamble, Ms. Kaley McCluskey, Dr. Ryan Mullins, Ms. Sahar Sahebdivan and Dr. Syman Stevens for their friendship, involvement and support.

Fifthly, I should thank a number of people who might have borne the burden of one kind or another, as a result of my doctoral activities, and perhaps particularly during the difficult and uncertain time of my transition from St Andrews to the Weizmann—a period rife with problems and misunderstandings. I certainly kept the travel office busy. I am told I spent far too much of my research money on *books*. And I doubtless caused my secondary supervisor, Dr. Natalia Korolkova, whose pastoral concern was appreciated in the first half of my Ph.D., more than her fair share of consternation during the second. Still, I trust no eggs were irreparably broken during the course of making this rather unusual omelette.

Sixthly, I must thank multifarious people for kind and practical help during the course of my studies: my Aunt Catherine and Uncle Jim Burns, for instance, for putting me up for the night in their 'sparest of spare rooms' whenever I was passing through London; Jana Silberg, for assisting me in my struggles with Israeli bureaucracy. (I do not know what you said or what I signed in the bank, but it seemed to do the trick). I am grateful to Stephen and Gillian Gamble for friendship, a room when I needed it, and the courage to tell me when I was being a 'belter'.

I am also indebted to David and Carol Pileggi, for opening their home in Jerusalem to a 'stranger in a strange land'. If a cup of water offered in kindness merits divine favour, whosoever opens their wine cellar to a displaced Ph.D. student must receive an eternal reward. My gratitude similarly extends to Stefano Recanatesi and his flatmates—David Lara, Luca Moleri and Pedro Astiaso—for frequent hospitality, for storing my chattels, and for the communion of good food and drink in 'a dry and thirsty land', to sweeten the last days of my sojourn in the Middle East. Indeed, there are many friends who have assisted me in some capacity during the course of my doctoral studies, whom time and space have prevented from being enumerated here, but I trust you know who you are.

Finally, I should thank my two families who have supported me during the course of my studies: my 'second family', which made a home for me in St Andrews, and in doing so transformed it—Mark, Jenny, Hamish, Sandy, Gregor and *even* Catriona (what home would be complete without a little sibling rivalry?); and my 'first family'—my mother, my sister Laura and my brother John, who time and again provided a peaceful half-way house on my travels, and prayerful support during all my adventures and endeavours. This thesis is dedicated to my father, who rejoiced to see this day: he saw it, and was glad.

# Contents

# About the Author

Photo by Gillian Gamble

**William Simpson** is a postdoctoral fellow at The Weizmann Institute of Science. He obtained a first in Theoretical Physics and Mathematics in 2010 and completed his Ph.D. at the University of St Andrews in 2014, having been awarded a SUPA Prize Studentship for doctoral studies. During the course of his research on Casimir forces, supervised by Prof. Ulf Leonhardt, William was a visiting student at The University of Trento and The Weizmann Institute of Science. He also founded a Physics and Philosophy Society (PhysPhil), and was the chief organiser of St Andrews' first two Philosophy of Physics conferences. William was the chairman and organiser of the Foundations of Casimir Physics symposium in Stockholm in 2013, and is a contributing co-editor of the book *Forces of the Quantum Vacuum*[5], which arose out of this seminal meeting. He has published in physics, philosophy and theology.

## Select Publications in Theoretical Physics

1. Forces of the Quantum Vacuum, Eds. Simpson, W. M. R., Leonhardt, U., World Scientific Publishing (early 2015) (Book)
2. The cutoff dependence of the Casimir force, Horsley, S. A. R., Simpson, W. M. R., Phys. Rev. A, 88, 013833 (2013) (Paper)
3. Divergence of Casimir stress in inhomogeneous media, Simpson, W. M. R., Horsley, S. A. R., Leonhardt, U., Phys. Rev. A, 87, 043806 (2013) (Paper)

[5] To be published by World Scientific Publishing in 2014/5.

4. The Casimir force in a compressive transformation medium, Simpson, W. M. R., Phys. Rev. A, 88, 063852 (2013) (Paper)
5. Exact solution for the Casimir stress in a spherically symmetric medium, Leonhardt, U., Simpson, W. M. R., Phys. Rev. D, 84, 081701(R) (2011) (Paper)
6. Ontological aspects of the Casimir Effect, Simpson, W.M.R., Studies in Hist. and Phil. Sci. Part B: Studies in Hist. and Phil. Mod. Phys. 2014, DOI: 10.1016/j.shpsb.2014.08.001 (Paper)

# Part I
# Foundations of Casimir Theory

# Chapter 1
# Introduction

*I mentioned my results to Niels Bohr, during a walk. "That is nice", he said, "That is something new". He mumbled something about zero-point energy. That was all, but it put me on a new track.*

Hendrik Casimir, *Comments on Modern Physics 5–6: 175*

## 1.1 Casimir's Thought-Experiment

The Casimir force derives its name from the scientist who first predicted it [1]. Hendrik Casimir was a Dutch physicist of the last century who made many contributions to Physics, but is most famous for predicting the phenomenon that carries his name. Casimir speculated that two perfect mirrors facing each other in a vacuum would experience an attractive force as a result of vacuum fluctuations present in the cavity, even at zero temperature—a prediction that was subsequently confirmed in a number of critical experiments decades later [2–5].

### *1.1.1 The Ground State Field Energy*

Casimir's thought-experiment concerns a cavity formed by two perfectly conducting, infinitely large plates, and the ground-state energy of the electromagnetic field in the cavity. In general, this 'zero-point energy' can be stated as a simple mode summation[1]:

$$E_{cavity} = \frac{\hbar}{2} \sum_{k} \rho(k)\, \omega_k, \tag{1.1.1}$$

[1] It is assumed that the reader is familiar with the quantisation of a simple harmonic oscillator and the notion of a zero-point energy in this context. For readers unversed in QED, its simple application to the quantised field should be taken on faith, along with a few subsequent appeals to field-theoretic notions, until the next chapter: *credo ut intelligam*.



W.M.R. Simpson, *Surprises in Theoretical Casimir Physics*, Springer Theses,
DOI 10.1007/978-3-319-09315-4_1

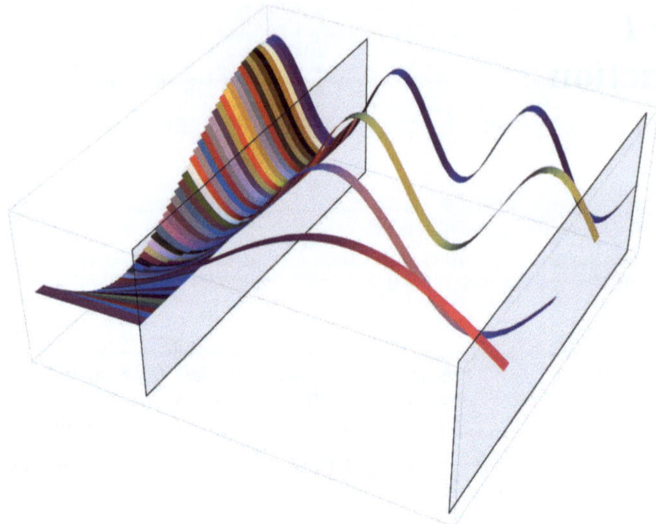

**Fig. 1.1** The walls of the cavity impose boundary conditions on the field modes; only modes with nodes at the locations of the boundaries can be sustained within the cavity

where $\rho(k)$ is the degeneracy of the mode with wave number $k$ and frequency $\omega_k$ satisfying Maxwell's equations and the relevant boundary conditions. For the purpose of this calculation, we position the plates at $x = 0$ and $x = a$, and take the height and breadth of the cavity to be $L_y$ and $L_z$. Because the plates are perfectly conducting, the transversal components of the electric field and the normal component of the magnetic field must vanish on these boundaries [6]:

$$E_y(x = 0) = E_y(x = a) = 0, \tag{1.1.2}$$

$$E_z(x = 0) = E_z(x = a) = 0. \tag{1.1.3}$$

This leads us to consider a different set of field modes from the plane waves of free-space. The field in the cavity is constituted by a set of discrete standing wave modes (see Fig. 1.1) with wave numbers

$$k_x = \frac{m\pi}{a}, \; k_y = \frac{p\pi}{L_y}, \; k_z = \frac{q\pi}{L_z}, \qquad m, p, q \in \mathbb{Z}^+. \tag{1.1.4}$$

Maxwell's equations are satisfied by the dispersion relation

$$\omega = ck = c\sqrt{k_x^2 + k_y^2 + k_z^2} = w_{mpq}, \tag{1.1.5}$$

where $w_{mpq}$ are the discrete eigenfrequencies of the set of standing waves between the plates. The electromagnetic field admits two polarisations, except for cases in which $m$, $p$ or $q$ is zero, for which there is only one. However, both polarisations are degenerate here,[2] so that

$$\rho(k) \rightarrow \rho_{mpq} = \begin{cases} 2 & m, p, q > 0, \\ 1 & m = 0,\ p = 0, \text{ or } q = 0. \end{cases} \tag{1.1.6}$$

The ground-state energy (1.1.1) of the system is summed over $m$, $p$ and $q$:

$$E_{cavity} = \frac{1}{2} \sum_{m,p,q}^{\infty} \hbar \rho_{mpq}\, \omega_{mpq}. \tag{1.1.7}$$

In the limit as $L_y$, $L_z \rightarrow \infty$, $k_y$ and $k_z$ become infinitesimal, and we can replace the $p$ and $q$ summations with integrals by setting

$$\mathrm{d}p = \frac{L_y}{\pi}\mathrm{d}k_y, \quad \mathrm{d}q = \frac{L_z}{\pi}\mathrm{d}k_z. \tag{1.1.8}$$

The triple sum in (1.1.7) can then be rewritten

$$\frac{cL_yL_z}{\pi^2} \sum_m \rho_m \int_0^\infty \mathrm{d}k_y \int_0^\infty \mathrm{d}k_z \sqrt{k_x^2 + k_y^2 + k_z^2}, \tag{1.1.9}$$

where the original weighting function has been replaced[3] by

$$\rho_m = \begin{cases} 2 & m > 0, \\ 1 & m = 0. \end{cases} \tag{1.1.10}$$

We can rewrite the double-integral above more succinctly in polar coordinates by replacing $k_y^2 + k_z^2$ with

$$k_\parallel^2 = k_y^2 + k_z^2, \quad k_y = k_\parallel \cos\phi, \quad k_z = k_\parallel \sin\phi, \tag{1.1.11}$$

so that

$$\int_0^\infty \mathrm{d}k_y \int_0^\infty \mathrm{d}k_z \sqrt{k_x^2 + k_y^2 + k_z^2} = \int_0^{\pi/2} \mathrm{d}\phi \int_0^\infty \mathrm{d}k_\parallel\, k_\parallel \sqrt{k_x^2 + k_\parallel^2}. \tag{1.1.12}$$

---

[2] This is not always the case. For example, in a cavity containing an inhomogeneous liquid the modes of the field are not degenerate. See Chap. 4.

[3] The points $p = 0, q = 0$ now have only *infinitesimal* weightings in the integral summation, so the factors of $1/2$ can be neglected.

We thus arrive at an expression for the ground-state energy of a cavity formed by two large, perfect mirrors, separated by a distance $a$:

$$\frac{E}{A} = \frac{\hbar c}{4\pi} \sum_m \rho_m \int_0^\infty k_\parallel \mathrm{d}k_\parallel \sqrt{k_x^2 + k_\parallel^2}. \tag{1.1.13}$$

In Eq. (1.1.13) $A = L_y L_z$ is equal to the area of one of the plates.

### 1.1.2 Regularising the Energy

Unfortunately this expression (1.1.13) for the energy of the cavity is infinite. However, infinite series abound in quantum field theory, and in this case a regularisation of (1.1.13) can be effected with a little physical common sense: Casimir asks us to recall that a perfect mirror is an idealisation that does not exist in nature; the cavity cannot keep arbitrarily high frequencies from leaking out [1]. To incorporate this physical fact about the dispersive nature of the mirrors, we can multiply (1.1.13) with a damping function $\exp(-\xi\omega/c)$, writing

$$\frac{\tilde{E}}{A} = \frac{\hbar c}{4\pi} {\sum_m}' \int_0^\infty k_\parallel \mathrm{d}k_\parallel \left(k_x^2 + k_\parallel^2\right)^{1/2} \exp(-\xi\omega/c), \tag{1.1.14}$$

where $\xi$ is a cut-off parameter and $\sum'_m = \sum_m \rho_m$. This sum is now clearly convergent[4]: as $\omega$ becomes large the contribution to the energy vanishes. However, this modified expression for the energy may no longer be considered as the total energy of the system of field plus mirrors. Taken literally as such, (1.1.14) would imply that the eigenfrequencies of the field eventually all tend to zero, an assumption for which there is no obvious motivation. Instead, (1.1.14) should be interpreted as the *part* of the total energy (1.1.13) that is associated with the configuration of the mirrors in the cavity; it is the free energy available to do work on the mirrors.

To assist with the calculation, we can reconceive our problem as a Casimir piston involving *three* parallel plates, in which the two outer plates are fixed and only the inner plate is permitted to move (see Fig. 1.2). We then need not worry about how the energy density varies outside the cavity [7]. Equation (1.1.14) is now supplanted by

$$\frac{\tilde{E}}{A} = -\frac{\hbar c}{4\pi} {\sum_m}' \frac{\partial}{\partial \xi} \int_0^\infty k_\parallel \mathrm{d}k_\parallel \exp\left(-\xi \sqrt{\left(\frac{m\pi}{a}\right)^2 + k_\parallel^2}\right) + \{a \to L - a\}, \tag{1.1.15}$$

where $a \to (L - a)$ indicates a repetition of the previous expression with $a$ replaced everywhere by $L - a$, to include the contribution from the right side of the cavity,

[4] Introducing an exponential within the summation over frequencies is a standard technique [9].

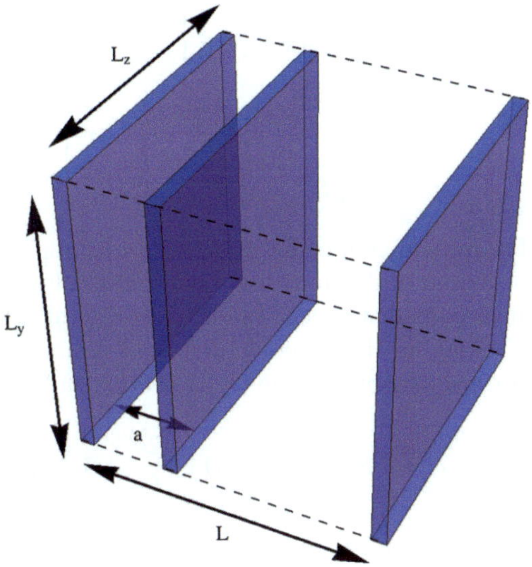


**Fig. 1.2** Schematic of the Casimir piston. Two fixed mirrors are positioned at $x = 0$ and $x = L$, enclosed by reflecting walls at $y = \pm L_y/2$ and $z = \pm L_z/2$ (*dashed lines*). Within the chamber is vacuum, and a moveable mirror at $x = a$

and the bracketed prefactor in (1.1.14) has been replaced by a partial derivative with respect to $\xi$. The regularisation may now proceed. The integral in (1.1.15) may be solved by applying the identity

$$\frac{d}{dk_{\parallel}}\left[\left(\xi^{-1}\sqrt{\left(\frac{m\pi}{a}\right)^2 + k_{\parallel}^2} + \xi^{-2}\right)e^{-\xi\sqrt{\left(\frac{m\pi}{a}\right)^2+k_{\parallel}^2}}\right] = -k_{\parallel}e^{-\xi\sqrt{\left(\frac{m\pi}{a}\right)^2+k_{\parallel}^2}}. \tag{1.1.16}$$

It follows from (1.1.15) and (1.1.16) that

$$\frac{E}{A} = -\frac{\hbar c}{4\pi}\sum_{m=0}^{\infty}{}' \frac{\partial}{\partial\xi}\left[\left(\frac{1+\xi m\pi/a}{\xi^2}\right)\mathrm{e}^{-\xi m\pi/a}\right] + \{a \to L-a\}. \tag{1.1.17}$$

Calculating the derivative with respect to $\xi$, then replacing prefactors by derivatives with respect to $\xi$, we obtain

$$\frac{\tilde{E}}{A} = \frac{\hbar c}{2\pi}\left[\frac{1}{\xi^3} - \frac{1}{\xi^2}\frac{d}{d\xi} + \frac{1}{2\xi}\frac{d^2}{d\xi^2}\right]\sum_{m=0}^{\infty}{}' e^{-\xi m\pi/a} + \{a \to L-a\}. \tag{1.1.18}$$

The $m$-dependent weighting on the primed sum may be removed by multiplying it by a factor of two and subtracting a term $\epsilon$ that does not depend upon the position of the mirror.[5] The summation within (1.1.18) is then a geometric series which can be evaluated, $\sum_{m=0}^{\infty} e^{-\alpha m} = 1/(1 - e^{-\alpha})$, yielding

$$\frac{\tilde{E}}{A} + \epsilon = \frac{\hbar c}{2\pi}\left[\frac{1}{\xi^3} - \frac{1}{\xi^2}\frac{d}{d\xi} + \frac{1}{2\xi}\frac{d^2}{d\xi^2}\right] e^{\xi\pi/2a}\operatorname{cosech}(\xi\pi/2a) + \{a \to L - a\}. \tag{1.1.19}$$

The introduction of the damping factor represents a rather artificial model for the behaviour of the mirrors at high frequencies. We therefore separate the energy into those parts that depend on the cut-off term $\xi$, and those that do not. In anticipation of taking the limit $\xi \to 0$ (that is, of removing the cut-off) the quantity to the right of the square brackets in (1.1.19) is expanded up to the third power in $\xi^3$:

$$e^{\xi\pi/2a}\operatorname{cosech}(\xi\pi/2a) \sim \frac{2a}{\xi\pi} + 1 + \frac{1}{3}\left(\frac{\xi\pi}{2a}\right) - \frac{1}{45}\left(\frac{\xi\pi}{2a}\right)^3. \tag{1.1.20}$$

Inserting (1.1.20) into (1.1.19) produces finally

$$\frac{\tilde{E}}{A} = \hbar c\left[\frac{3L}{\pi^2\xi^4} - \frac{1}{\pi\xi^3} - \frac{\pi^2}{720a^3} - \frac{\pi^2}{720(L-a)^3}\right]. \tag{1.1.21}$$

As expected, the energy becomes increasingly large in the limit where the effects of the regularisation disappear – that is, as $\xi \to 0$. However, by expanding in terms of the cut-off parameter $\xi$, we have been able both to identity and distinguish a diverging part of the energy and a part that depends on the position of the mirror. Remarkably, the latter part is independent of $\xi$. One may interpret this to mean that, so long as $\xi/a$ is negligibly small (i.e. neglecting the positive powers of $\xi$ in (1.1.21) is legitimate), the part of the energy that depends on $a$ is independent of how the mirror becomes transparent at high frequencies/wave–vectors.

### *1.1.3 The Finite Casimir Force*

We determine the force on the moveable mirror by differentiating the energy (1.1.21) with respect to its position variable $a$. The first two terms depending on $\xi$ disappear, effectively completing the removal of the cut-off. The force on the moveable mirror is then found to be

$$\frac{F}{A} = -\frac{\pi^2\hbar c}{240a^4} + \frac{\pi^2\hbar c}{240(-a+L)^4}, \tag{1.1.22}$$

[5] The term $\epsilon$ corresponds to one half of the contribution of the sum at $m = 0$ for both sides of the cavity. It is equal to $\hbar c/\pi\xi^3$.

which consists of an attractive force directed towards the parallel plate on the left, and an attractive force in the opposite direction, pointing towards the plate on the right. In the limit as $L \to \infty$ (that is, as the third plate is removed an infinite distance to the right), we recover the famous expression for the Casimir force between two parallel plates [1]:

$$\frac{F}{A} = -\frac{\pi^2 \hbar c}{240 a^4}. \tag{1.1.23}$$

It is worth taking a moment to reflect on this remarkable result. From the infinite ground-state energy of the quantised electromagnetic field, which had hitherto been dismissed by many physicists (up to the calibre of Dirac) as physically meaningless, Casimir discovered a finite and physical contribution that depends upon the size of the cavity, varying with the inverse fourth power of the distance between the plates. It must be emphasised that this force is tiny: for two $1 \times 1$ cm plates separated by 1 $\mu$m, Casimir's equation predicts a force of 0.013 dyne, comparable to about 1/1,000 the weight of an ordinary housefly [8]. Nevertheless, there is an energy available for doing mechanical work.

Furthermore, the expression derived is surprisingly simple. The Casimir force, thus formulated, appears to depend on nothing besides a mechanical property of the system (the distance $a$) and a prefactor incorporating Planck's constant $\hbar$ and the speed of light in a vacuum $c$. On this analysis, the phenomenon appears to arise solely due to a topological modification of the vacuum, effected in this case by the external boundary conditions imposed on the field by two parallel plates. In fact, things are somewhat more complicated. We will return to the subject of the nature of the Casimir force in Chap. 3.

### *1.1.4 Renormalisation and Background Energies*

Theorists sometimes refer to an additional step in the calculation of the Casimir force, omitted in the discussion above, in which the energy is *renormalised*. This renormalisation is typically effected prior to the removal of the regulariser, and it involves the computation and subtraction of a so-called 'background energy' that reproduces and therefore removes the diverging terms. The renormalised energy that remains is then a finite quantity. In our case, however, the divergences disappear in the course of computing the force derivatives, and the remaining terms involving the cut-off are nullified in the limit $\xi \to 0$. The finite energy associated with the Casimir Effect can then be computed retroactively.

Still, the purist may feel faintly troubled by this departure from the standard procedure, and there is a good reason why we have omitted it from the calculation of the force here: it does not entirely work. Let's try to follow the recipe exactly, without recourse to the usual mathematical bag of tricks, such as zeta functions or the Euler-Maclaurin series [9, 10]. Instead, we will only allow ourselves Casimir's physically motivated regularisation. The background energy $E_\infty$ (the energy of the free field) can be computed as follows. In general, the energy per unit volume is given by

$$\frac{E_\infty}{V} = \frac{\hbar c}{2} \int \rho(k) \frac{k \, \mathrm{d}k^3}{(2\pi)^3}, \tag{1.1.24}$$

where $\rho(k)$ is a density function. Here, $\rho(k) = 2$. The volume of free space we are considering is $V = L^3$. Therefore, the background energy per unit area can be conveniently written in the form

$$\frac{E_\infty}{A} = \frac{\hbar c L}{2\pi^2} \int_0^\infty \int_0^\infty k_\parallel \, \mathrm{d}k_\parallel \mathrm{d}k_x \sqrt{k_x^2 + k_\parallel^2}, \tag{1.1.25}$$

in which the volume integral was recast as a cylindrical polar integral, introducing $k_\parallel$ as before (1.1.11–1.1.12), and then integrating over the azimuth. This quantity is clearly infinite. In order for the subtraction between the two energies to be well-defined, renormalising the energy $E \to \tilde{E}$, both must be submitted to the same regularisation:

$$\tilde{E} = \lim_{\xi \to 0} \left[ E(\xi) - E_\infty(\xi) \right]. \tag{1.1.26}$$

The regularised background energy, pursued in the same spirit as before, takes the form

$$\frac{\tilde{E}_\infty}{A} = -\frac{\hbar c L}{2\pi^2} \frac{\partial}{\partial \xi} \int_0^\infty \int_0^\infty k_\parallel \, \mathrm{d}k_\parallel \mathrm{d}k_x \exp\left( -\xi \sqrt{k_x^2 + k_\parallel^2} \right). \tag{1.1.27}$$

The integral over $k_\parallel$ is performed, once again using the identity (1.1.16), leading to

$$\frac{\tilde{E}_\infty}{A} = -\frac{\hbar c L}{2\pi^2} \frac{\partial}{\partial \xi} \int_0^\infty \mathrm{d}k_x \left( \frac{1 + \xi k_x}{\xi^2} \right) e^{-\xi k_x}. \tag{1.1.28}$$

Calculating the derivative with respect to $\xi$, then rewriting prefactors as derivatives with respect to $\xi$, we find that

$$\frac{\tilde{E}_\infty}{A} = \frac{\hbar c L}{\pi^2} \left( \frac{1}{\xi^3} - \frac{1}{\xi^2} \frac{\partial}{\partial \xi} + \frac{1}{2\xi} \frac{\partial^2}{\partial \xi^2} \right) \int_0^\infty \mathrm{d}k_x \, e^{-\xi k_x}. \tag{1.1.29}$$

The integral is easily evaluated as $1/\xi$ and (1.1.29) simplifies to

$$\frac{\tilde{E}_\infty}{A} = \hbar c \frac{3L}{\pi^2 \xi^4}. \tag{1.1.30}$$

This is identical to the first term of (1.1.21). It indicates that this divergence in the energy is an artifact of our treatment of empty space. In the absence of cavity walls (or, for cavity walls that are infinitely separated from each other), the Casimir energy should be zero. The subtraction of this term from any Casimir calculation in a Minkowsi space appears entirely justified.

However, there is a second divergent term in (1.1.21) which is neither removed by the standard renormalisation procedure (1.1.26) nor assigned semantics, producing an infinite contribution to the energy in the limit as $\xi \to 0$. This additional divergence is associated with the presence of the mirrors and the $m = 0$ contribution to the energy; it appears to correspond to waves propagating parallel to the plates. In the calculation of the Casimir force in a vacuum, it does not depend upon the position of the moveable mirror, and consequently makes no contribution to the force. However, this independence may be entirely fortuitous. We have no grounds for supposing that this will always be the case, and from what follows in subsequent chapters, we may find reason for doubting that it *is*. As we will see, even in more sophisticated theories of the Casimir Effect, divergences seem to lie in wait along these axes. But first, we should place this discussion in its proper context.

## 1.2 Dispersion Forces

It was Bohr, in fact, who put Casimir on the right track for his famous prediction by mumbling "something about zero-point energy" during one of their conversations. Casimir was working on colloids[6] at the time in connection with the Philips Research Laboratory in the Netherlands. The properties of these viscous materials are determined by van der Waals forces. However, the theory developed by Fritz London [11, 12] failed to explain the experimental measurements for the interaction between two molecules [13], until Casimir and his colleague Dirk Polder corrected it [14]: the assumption that one dipole moment polarizes another particle instantaneously is only valid if the distance is much smaller than the typical wavelength of the fluctuating fields between the two molecules. On abandoning this so-called *quasi-static approximation*, the predicted force is significantly modified.

Casimir realised that the problem could be solved more simply in terms of vacuum fluctuations, using normal-mode quantum electrodynamics. He then considered what might happen if the two molecules were replaced by two mirrors facing each other in a vacuum, leading to the famous prediction we have just re-derived [1]. The Casimir force, in fact, is part of a family of *dispersion forces* that Casimir studied, along with other physicists, which we will briefly review in this section. They are forces that share a number of common features:

First, they are regarded as *effective* electromagnetic forces; they are modelled as forces acting on each object as a whole, in a system in which the interacting objects are spatially separated and specifiable in terms of their centre-of-mass positions and orientations, ensuring no overlap in their respective wave functions. Secondly, the

[6] A colloid is a mixture in which particles of one component are suspended in a continuous phase of another component.

dispersive characterisation of these forces alludes to the variation with frequency of the relevent atomic properties (or the properties of the media) involved in producing them. Both the polarisability of atoms, and the permittivities and permeabilities of bulk materials, exhibit a frequency-dependence that more realistic models attempt to capture. The study of dispersion forces, in the past, was typically restricted to material objects and fields interacting in their respective ground-states, but many results have now been generalised to include thermal fields and systems in excited states. In this study, however, we wish to focus on purely quantum-mechanical effects, and will restrict ourselves to the consideration of systems at zero temperature in their ground-states.

Two common ways of classifying dispersion forces are by making reference either to the characteristic length scale of the interacting system (how far apart the interacting objects are) or the size of the objects themselves and how they are modelled (as microscopic particles or bulk media). The first division falls between *retarded* and *non-retarded forces*. As Verwey and Overbeek discovered [13], when the distance between the interacting objects is on the scale of atomic transition wavelengths, the force is no longer essentially the electrostatic Coulomb interaction that dominated the non-retarded case. For these retarded forces, the finite speed of light and hence the retardation of the field must be taken into account. The second classification is a division that cuts three ways across two length scales, partitioning dispersion phenomena into van der Waals, Casimir-Polder and Casimir-Lifshitz forces. It is worth taking a few moments to touch on each of these categories.[7]

## *1.2.1 Van der Waals Forces*

The van der Waals rubric is typically restricted to the microscopic case of forces between atoms and molecules. These forces are ubiquitous throughout nature. In physical chemistry they feature in the binding of atoms to form molecules [15, 16], and contribute to the total binding energy of liquids and solids, with concomitant effect upon their macroscopic properties [17]. In biology, van der Waals forces are significant in the interaction of molecules within living cells, affecting material transport through cell membranes [18, 19]. Even on the scale of astrophysics and cosmology, van der Waals forces are postulated a role in planet formation, assisting in the creation of dust grains and their growth into planetesimals (after which, gravitational forces take over) [20, 21]. It is speculated that dispersion forces may be the dominant binding influence in asteroids [22].

We have already alluded to the original context of Casimir's seminal work in this field. In the DLVO theory of colloid behaviour, comprising Derjaguin's and Landau's

[7] For more detailed reviews of van der Waals and Casimir-Polder phenomena, which this section is especially indebted to, see [25, 48]. Also, Ephraim Shahmoon has recently written a more detailed introduction to van der Waals and Casimir-Polder phenomena for our new book, Forces of the Quantum Vacuum: an introduction to Casimir Physics (World Scientific Publishing, to be released in 2015).

theory of colloidal dispersions in 1941 [23], and the corroborative results of Verwey and Overbeek arrived at 7 years later [13], the stability of colloidal suspensions was accounted for by strong but short-ranged van der Waals attractions countered by stabilizing electrostatic repulsions. The colloidal particles are subject to both kinds of forces: the former arising from spontaneous polarisations, and tending to cluster the particles together; the latter resulting from a sheath of counter-ions accumulated on the particle surfaces, and tending to repel the particles apart.

#### 1.2.1.1 Eisenschitz and London's Theory

For interactions over small length scales, van der Waals forces favour an electrostatic approach. The force is dominated by the Coulomb interaction of the particles and the system can be modelled by a dipole-dipole interaction Hamiltonian. In this case, both atoms are unpolarised and all orientations of each dipole are equally likely. By applying second-order perturbation theory, in which the energy of a system of two uncoupled ground-state atoms is shifted by the dipole-dipole Hamiltonian, Eisenschitz and London were able to show that, whilst the first-order contribution vanishes, the second-order term does not [11, 12]. This second-order energy shift was identified as the London potential [12],

$$U(r) = -\frac{1}{24\pi^2\epsilon_0^2|\mathbf{r}_1 - \mathbf{r}_2|^6}\sum_{k,l}\frac{|\mathbf{d}_{0k}^1|^2|\mathbf{d}_{0l}^2|^2}{E_k^1 - E_0^1 + E_l^2 - E_0^2} \propto -\frac{1}{r^6}, \tag{1.2.1}$$

where $\mathbf{r}_1$ and $\mathbf{r}_2$ are the positions of the atoms, $r$ is the distance between them, the sum is over virtual transitions from the ground states (with energies $E_0^1$, $E_0^2$) to excited states $l, k$ (with energies $E_k^1$, $E_l^2$), and $\mathbf{d}_{0k}^1$, $\mathbf{d}_{0l}^2$ are the dipole matrix elements of the first and second atoms respectively. The potential is attractive and propotional to $1/r^6$, and may be interpreted as resulting from fluctuations of the charge distribution that couple directly via the electrostatic Coulomb interaction.

#### 1.2.1.2 Casimir and Polder's Theory

As Overbeek suspected [13], beyond a certain length scale Eisenschitz and London's approach begins to break down. Attempting to fix it on its own terms proved rather difficult, however. Instead, Casimir and Polder opted to use normal mode quantum electrodynamics (QED) [14]. This involved studying the effects of introducing into the Hamiltonian an operator for the interaction between a neutral atom and the radiation field, in recognition of the fact that the fluctuations of the transverse electromagnetic field also contribute to the potential. In addition to the Coulomb interaction, the atoms influence each other indirectly through the field. Using fourth-order perturbation theory the correctional energy shift of introducing the interaction to a system

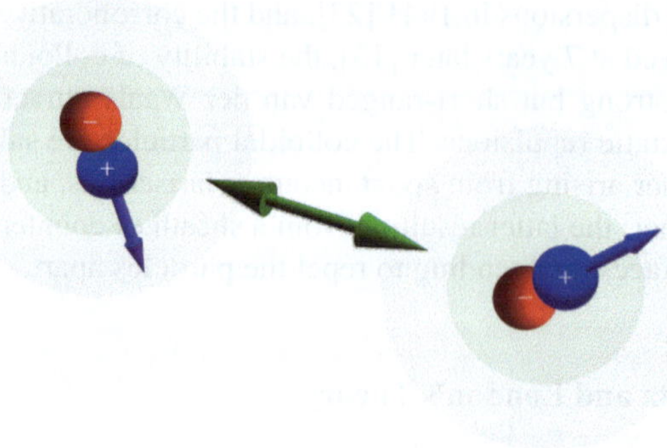

**Fig. 1.3** The van der Waals force, as conceived in Eisenschitz and London's theory, is an electrostatic interaction between spontaneously polarised particles, dominated by the Coulomb force (*green arrow*). The particles are modelled as dipoles with arbitrary dipole moments (*blue arrows*)

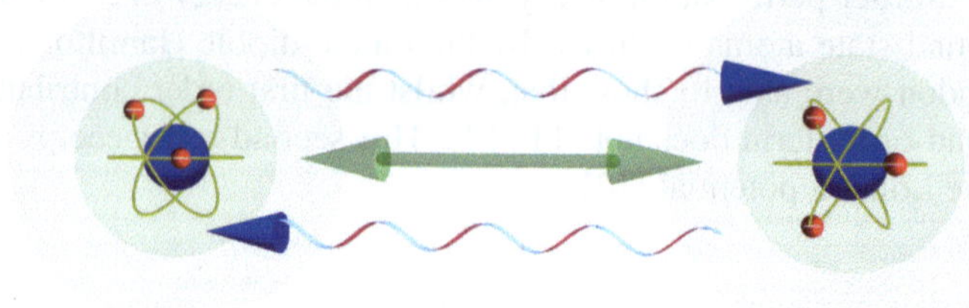

**Fig. 1.4** The van der Waals force, as conceived in Casimir and Polder's theory. Here, the interacting objects are modelled as fluctuating quantum particles coupled to the electromagnetic field. The electrostatic interaction (*green arrow*) is supplemented by an interaction via the fields (*blue arrows*)

comprised of uncoupled ground-state atoms, and a field in the vacuum state, lead to a somewhat more complicated expression for the potential, but one in which the London result appears as good approximation in the non-retarded limit. However, for the strongly retarded limit, the van der Waals potential yields [14] (Fig. 1.3)

$$U(r) \propto -\frac{\hbar c \alpha_1 \alpha_2}{\pi^3 \epsilon_0^2} \frac{1}{r^7}, \tag{1.2.2}$$

where $\alpha_1$ and $\alpha_2$ are the polarisabilities of the two atoms. Significantly, this potential falls off more rapidly with the distance between the atoms than the London potential, being proportional to $1/r^7$, as the experimental evidence suggested (Fig. 1.4).

#### 1.2.1.3 Modifications of van der Waals Forces

As intimated earlier, such methods have proven extendible to cases involving finite temperatures, excited atoms, atoms with magnetic and chiral properties, and atoms situated in non-trivial environments. van der Waals forces can also be modified by the presence of nearby bodies (e.g. plates, spheres, planar cavities). Most of these scenarios are beyond the concerns of this study. However, for a basic example of environment-dependence that is of some relevance, we can imagine placing the interacting components within a medium of permittivity $\epsilon$. The potential for this case has been determined: it is modified to [24]

$$U_\epsilon(r) = \frac{C_F}{n\epsilon^2} U(r), \tag{1.2.3}$$

where $U(r)$ is the original potential given by (1.2.2), which is modified by a constant consisting of the refractive index $n = \sqrt{\epsilon}$, a screening effect $1/\epsilon^2$ that is due to the polarisation of the medium, and a final factor $C_F = (3\epsilon/(2\epsilon + 1))^4$ that is the result of local field corrections [25].

#### 1.2.1.4 Measuring van der Waals Forces

Accurate measurements of van der Waals forces have been achieved using scattering experiments. For example, the attenuation of an atomic beam due to van der Waals forces, as a result of being passed through a stationary gas, has been used to infer a potential that can be checked against predictions, confirming the $1/r^6$ non-retarded behaviour, e.g. [26]. In more sophisticated experiments, two such beams have been crossed, leading to a scattering of atoms in the region of their intersection, e.g. [27]. The scattered atoms are counted as a function of the scattering angle, and the potential is then deduced.

### *1.2.2 Casimir-Polder Forces*

A second classification of dispersion forces is the Casimir-Polder force, which arises between an atom or molecule and a solid body.[8] In this case, the two objects of concern are the microscopic particle and the medium of the solid body. The theory of Casimir-Polder forces has been applied to various phenomena, such as the adsorption of a single atom or molecule to a surface [28], and fruitfully employed in technological applications, such as the atomic force microscope [29], whose diagnostic utility is predicated upon the material-dependent variation of the force between a probe particle and the surface being probed. Again, there are different theoretical approaches to this type of problem.

[8] This section is especially indebted to the discussion in [25].

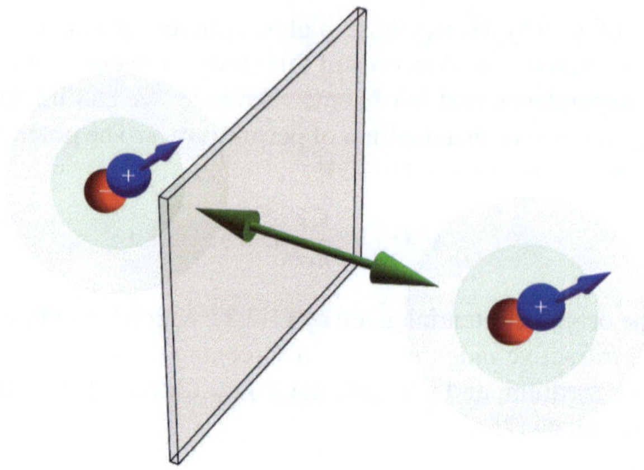


**Fig. 1.5** The Casimir-Polder force, as conceived in Lennard and Jones' theory, is a simple modification of an electrostatic van der Waals problem in which the effect of the solid body (*grey cuboid*) on the particle (*right*) is modelled as an image dipole (*left*) of the same orientation

#### 1.2.2.1 Lennard-Jones' Theory

In parallel with Eisenschitz and London's approach to van der Waals problems, the Lennard-Jones theory of the Casimir-Polder force uses the model of two dipoles, deriving the force from the electrostatic interaction between them [30]. The effect of the medium is incorporated by implementing the method of images: the atom's fluctuating dipole induces an image dipole behind the surface of the body, which acts as a second atom. However, in this case the orientation of the two dipoles can no longer be treated as independent. The average interaction energy is found to be [30]

$$U(r) = -\frac{1}{4\pi\epsilon_0}\frac{1}{12r^3}\sum_n |\mathbf{d}_n|^2 \propto -\frac{1}{r^3}, \tag{1.2.4}$$

where $\{\mathbf{d}_n^2\}$ are the dipole matrix elements of the single particle, and $r$ is the distance between the particle and the surface of the body. This average energy is identified as the Casimir-Polder potential. On Lennard-Jones' approach, charge fluctuations in the particle induce charge fluctuations in the medium. This is a longer-range interaction compared to the van der Waals force, which is not altogether surprising; the mirror is comprised of many atoms and produces more scattering (Fig. 1.5).

#### 1.2.2.2 Casimir and Polder's Theory

This result was also generalised by Casimir and Polder using normal mode QED, incorporating the indirect interaction of the body and the atom through the transverse electromagnetic field [14]. In this case, as in the case of the Casimir force, the presence of the body is now registered as a boundary condition: the fluctuating electromagnetic field is forced to vanish at the surface of the body. The Lennard-Jones potential is recovered in the non-retarded limit. As the separation between the body and the particles is increased, however, the potential tends towards

$$U(r) \propto -\frac{\hbar c \alpha}{\pi^2 \epsilon_0} \frac{1}{r^4}, \tag{1.2.5}$$

where $\alpha$ is the polarisability of the atom, yielding a force that falls off more rapidly with the distance between the atom and the body than (1.2.2) (Fig. 1.6).

#### 1.2.2.3 Linear Response Theory

Both these approaches to Casimir-Polder potentials share a severe limitation, in common with our earlier calculation of the Casimir force: they treat the media involved as perfectly reflective, perfectly rectilinear boundaries, without reference to their material constituents.[9] However, real media are not homogeneous geometric blocks that respond uniformly to light at all frequencies. In the following chapter, we will explore a quantum mechanical theory of light in media (macrosopic QED) that is capable of addressing these physical facts.

For the moment, let us simply note on passing that a generalisation to bodies of arbitrary shapes and dielectric properties is possible. Dispersion forces depend on fluctuations that can be related to linear response functions via the fluctuation-dissipation theorem [31]. Using linear response theory, the Casimir-Polder potential can be written succinctly in the form [32–34]

$$U(\mathbf{r}) = \frac{\hbar \mu_0}{2\pi} \int_0^\infty \mathrm{d}\xi \, \xi^2 \alpha(i\xi) \, \mathrm{Tr}\left[\mathbf{G}_S(\mathbf{r}, \mathbf{r}, i\xi)\right], \tag{1.2.6}$$

where $\mathbf{G}_S$ is the scattering part of a Green function $\mathbf{G}(\mathbf{r}, \mathbf{r}')$ describing the propagation of an electric field from a dipole source at position $r'$ to the point of measurement $r$. The fluctuations of the atomic dipole are incorporated in terms of the atomic polarisability of the atom $\alpha$; the fluctuations of the field are proportional to the classical Green function of the electromagnetic field. The Green function incorporates the specific geometry of the problem, being dependent on a given arrangement of bodies. The integral (1.2.6) is performed over imaginary frequencies, for reasons which will become apparent in Chap. 2. Both the atomic polarisability $\alpha(i\xi)$ and the Green function of the field $\mathbf{G}_S(\mathbf{r}, \mathbf{r}, i\xi)$ are rewritten as functions of imaginary frequency. For the example of an unexcited atom close to a semi-infinite half-space

[9] In addition, the media and fields in these models are not treated as a self-consistently coupled quantum mechanical structure.

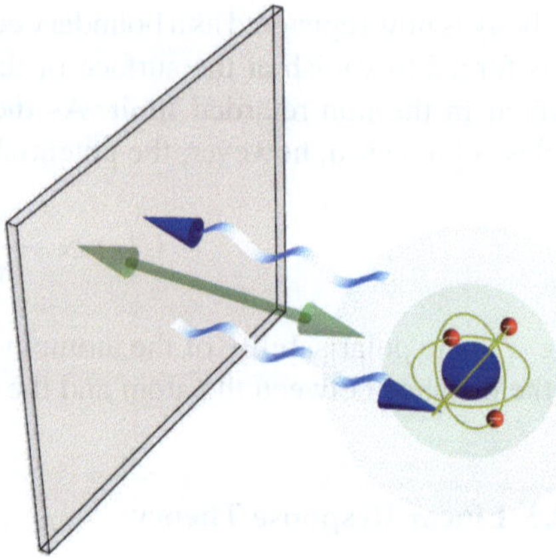

**Fig. 1.6** The Casimir-Polder force, as conceived in Casimir and Polder's theory, incorporates the interaction between the particle (*right*) with the body via the electromagnetic fields (*blue arrows*). The particle is now modelled as a fluctuating quantum atom, and the body is modelled as a perfect mirror (*the cuboid*)

of permittivity $\epsilon(\omega)$, which Lennard-Jones considered, the non-retarded potential is given by [34, 35]:

$$U(r) = -\frac{\hbar}{16\pi^2\epsilon_0 z^3}\int_0^\infty \mathrm{d}\xi\,\alpha(i\xi)\frac{\epsilon(i\xi)-1}{\epsilon(i\xi)+1} \propto -\frac{1}{r^3}. \tag{1.2.7}$$

It is attractive and propotional to $1/r^3$. Lennard-Jones original result is recovered in the limit $\epsilon \to \infty$, in which the medium is treated as a perfectly conducting mirror.

#### 1.2.2.4 Measuring Casimir-Polder Forces

Casimir-Polder forces can be measured in a similar way to van der Waals forces by determining the deflection or attenuation of atomic beams. In this case, however, the source of the interposition is neither a gas nor a second beam of particles, but a material body. For example, in a pioneering experiment Raskin and Kusch were able to confirm the predicted attractive behaviour of the non-retarded Casimir-Polder force in an arrangement using a cylinder as the deflecting body [36].

However, there are many alternative ways of measuring Casimir-Polder forces that have been developed. A comparatively recent technique involves the use of optical traps for confining atoms in an oscillatory motion (for an example, see [37]). The motion of the oscillation is modified in a measurable way by introducing a surface sufficiently close to interact with the atoms via the Casimir-Polder force. The change in the trapping potential is a function of the distance between the surface and the atoms confined in the trap. Casimir-Polder forces have also been measured in diffraction experiments in which the atomic beam is directed at a diffraction grating of appropriate width to diffract the matter wave associated with the beam [38]. The material surface presented by the walls of the slit subjects the beam to a

Casimir-Polder force, inducing a phase shift in the matter wave and modifying the interference pattern.

## *1.2.3 Casimir-Lifshitz Forces*

### 1.2.3.1 The Casimir and de Boer Forces in Contrast

Casimir's approach to computing dispersion force between macroscopic bodies was not the only contender. An alternative microphysical argument had been put forward in 1936 by Hamaker and Boer [39, 40]. By summing the microscopic London interactions of the atoms (idealised as spherical particles) contained in two bodies over their respective volumes, a different expression for the 'Casimir energy' was discovered:

$$U(r) = -\int_{V_1} \mathrm{d}^3 r_1 \rho_1 \int_{V_2} \mathrm{d}^3 r_2 \rho_2 \frac{C_{vw}}{r^6} \tag{1.2.8}$$

$$= -\int_{V_1} \mathrm{d}^3 r_1 \int_{V_2} \mathrm{d}^3 r_2 \frac{C_h}{\pi^2 r^6}, \tag{1.2.9}$$

where $C_h = \pi^2 \rho_1 \rho_2 C_{vw}$ is the Hamaker constant, $C_{vw}$ is the London van der Waals constant, and $\rho_1$ and $\rho_2$ denote the number densities of the atoms in the two bodies. This model lead to Boer's prediction in 1936 of an attractive pressure-force,

$$F_B = -\frac{C_h}{12\pi r^3}, \tag{1.2.10}$$

between two parallel dielectric half-spaces, separated by a distance $r$. Unlike the Casimir force, de Boer's expression varies with the *third* power of the distance between the two bodies.

### 1.2.3.2 Lifshitz Theory

As Buhmann points out [25], Casimir and de Boer's approaches to the problem are, in a sense, exact opposites: de Boer's force is non-retarded; Casimir's force is retarded; de Boer's force is caused by the fluctuations of material charge; Casimir's force is founded on the fluctuations of the electromagnetic field; de Boer's force is the result of a simple sum of two-body interactions that is only applicable to weakly dielectric bodies; Casimir's force assumes the infinitely strong dielectric properties of a perfect mirror.

The two results, in fact, turn out to be limiting cases of a more general apparatus for calculating dispersion forces between bodies: Lifshitz theory, valid for arbitrary distances and dielectric properties, recovers both contenders within their respective limits, offering a stochastic theory of the Casimir Effect that causally unifies the fluctuating fields between the interacting bodies with the fluctuating materal they are made of [41–43], introducing a random source of polarisation within the bulk description of the dielectric that is in accordance with the fluctuation-dissipation theorem [41]. We will discuss this approach in more detail in the following chapter.

Nevertheless, allowing for its retarded limit, the force predicted by Casimir's theory (1.1.23) is an idealisation that is not encountered in nature.[10] For instance, it requires the use of perfect mirrors, which do not exist, and it fails to take into account the presence of thermal fluctuations or dissipative processes in the materials. Before the Casimir force could be measured, such factors had to be taken into account, in some measure. Lifshitz' result, offering a more general expression for the Casimir force, has proven an important benchmark for the prediction of dispersion forces in more realistic cases, enjoying significant experimental verification [44, 45] (Fig. 1.7).

The unification effected by Lifshitz theory is arguably skin-deep, however, occurring at the stochastic and the phenomenological levels via the insertion of suitable correlation functions to recover the right magnitude of the electromagnetic field on average. We will return to the subject of the nature of the Casimir force, following a short but necessary review of macroscopic quantum electrodynamics.[11] But first, we should briefly acknowledge some of the pioneering experiments that have confirmed the presence of Casimir forces in our world.

## 1.3 Experimental Evidence

### *1.3.1 Early Measurements*

Casimir's prediction of an attractive quantum-mechanical force between two parallel mirrors remained a theoretical curiosity for some time. The first serious attempt to test Casimir's prediction in 1948 was led by Marcus Sparnaay 10 years later [3]. His equipment consisted of two parallel plates, the lower plate mounted on a heavy pedestal to suppress vibrations, the upper plate suspended on a spring obeying Hooke's law. The bottom plate could be moved by a screw system with a precision of less than 5 nm. By bringing it towards the upper plate, the extension of the spring was measured, demonstrating an increase in force with a decrease in separation distance. However, the errors in this experiment were enormous (some of the problems including the surface roughness of the plates, and the difficulty of

[10] This is worth emphasising. For instance, Jaffe's claim that the (idealised) Casimir formula (1.1.23) has been verified to 1 % [49] is somewhat misleading. See Chap. 3.

[11] Macroscopic quantum electrodynamics is discussed in Chap. 2. For further discussions about the nature of the Casimir force, see Chap. 3.

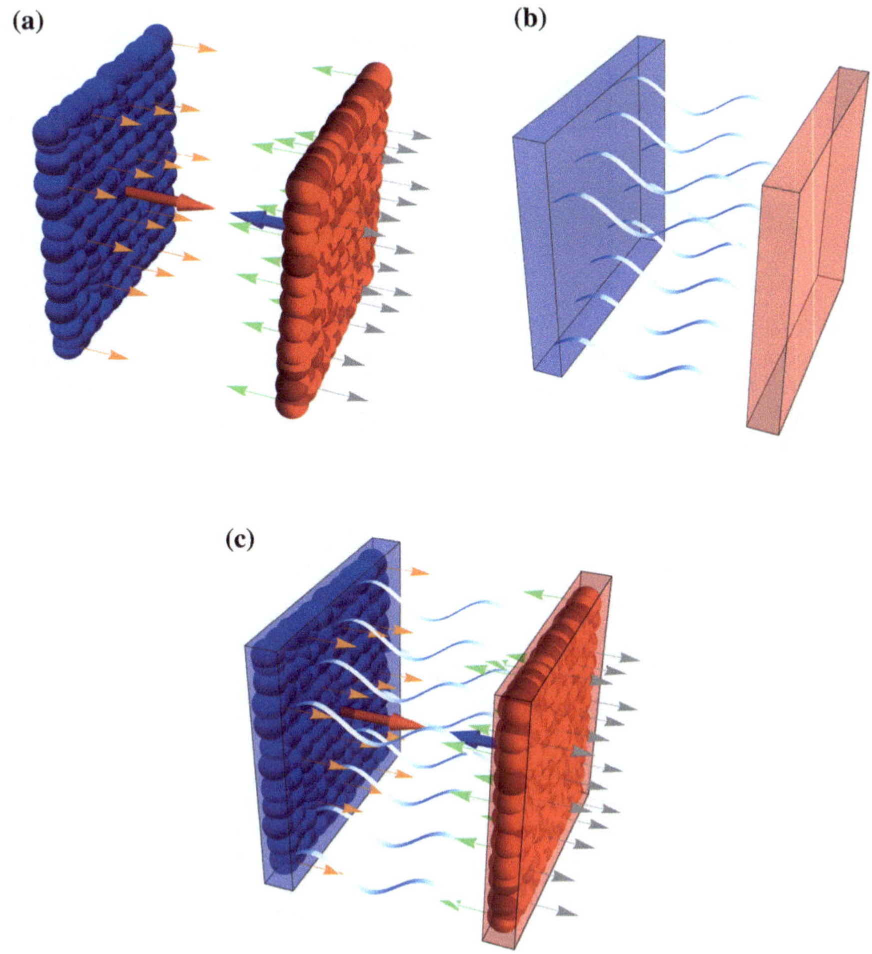


**Fig. 1.7** The Casimir Effect, as depicted by **a** Hamaker's theory: a sum over microscopic dipole-dipole interactions; **b** Normal mode QED: a sum over electromagnetic field modes between simple boundaries; **c** Lifshitz theory: a stochastic electromagnetic field created by fluctuating polarisations in a dielectric. In Lifshitz theory, the dielectric functions incorporate something of the microscopic detail of macroscopic bodies by virtue of their frequency-dependence

maintaining parallelism at the requisite accuracy) and the results could only be said to be 'consistent' with the Casimir effect; they could not unambiguously confirm its existence.

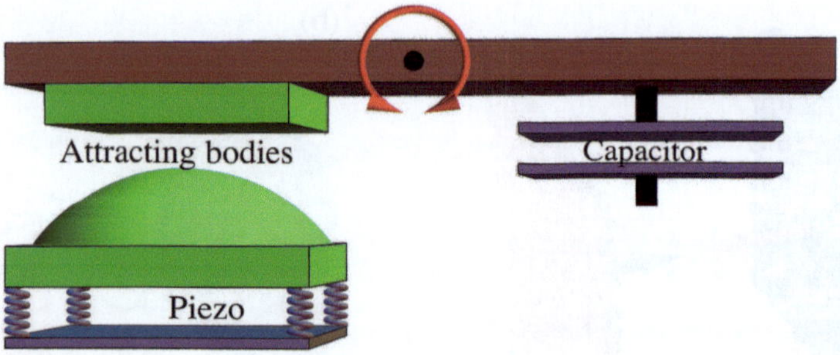


**Fig. 1.8** A torsion pendulum model. The beam is mounted from a thin wire and is free to rotate about the pivot. Two attractive surfaces (*green*) are brought into proximity by adjusting the height of the piezo-mounted body. The force required to maintain equilibrum is deduced from the voltage of a capacitance bridge

## *1.3.2 High Precision Measurements*

### 1.3.2.1 Lamoreaux's Experiment

It was not until 1978 that clearer experimental evidence for the effect was presented by van Blokland and Overbeek [4], who circumvented the problem of keeping the plates parallel by performing the measurement between a metallic plate and a metallic sphere. It is Lamoureaux in 1997, however, who is generally regarded as having supplied the more decisive evidence [2], inaugurating the era of high precision measurements. Lamoreaux adopted the same sphere-plate configuration of van Blokland and Overbeek, incorporating a torsional pendulum pivoted and connected to the plate (see Fig. 1.8). When the lens and plate were brought together to within several microns, the Casimir force pulled the two objects together, causing the pendulum to twist. Lamoreaux was able to measure the restoring force necessary to keep the pendulum angle fixed.

### 1.3.2.2 Atomic Force Microscopy

Shortly after Lamoreaux's ground-breaking experiment, a number of Casimir force measurements were made using variations on the atomic force microscope apparatus (see Fig. 1.9). In this alternative apparatus, a metal plate mounted on a piezoelectric translator interacts with a small metal sphere attached to a sensitive cantilever. As the two bodies are brought into proximity, the bending of the cantilever is detected by a laser beam reflected off the back of the cantilever, and observed as a change in the signal of a detector monitoring the difference in light intensity between the top and bottoms halves of the detector.

Whilst claims to have measured the Casimir force by this method with an accurracy of 1 % or better [5] may require cautious evaluation [46], this technique has been

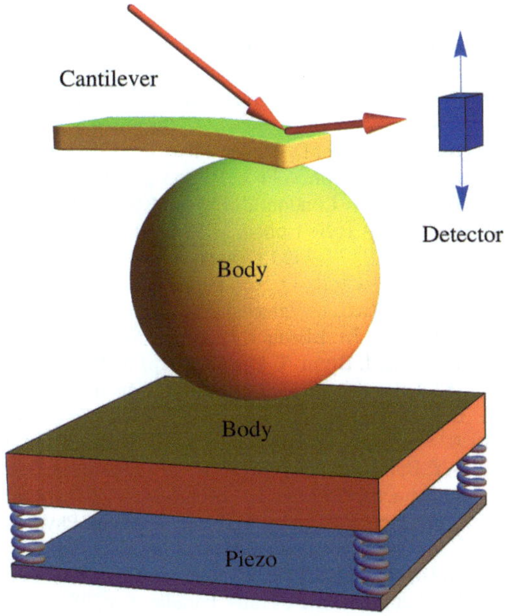


**Fig. 1.9** Atomic force microscope. A metal plate mounted on a piezoelectric translator interacts with a small metal sphere attached to a sensitive cantilever, bending the cantilever and displacing the reflection of a laser beam. The deflection of the beam is measured by a photodiode detector

fruitfully applied to a number of different manifestations of the Casimir Effect, including the effects of surface roughness on Casimir forces [47], and even the phenomenon of repulsive Casimir forces [42], which may arise between two objects when one is optically ‘thinner’ than the intervening medium. (We can regard the optically thinner medium as effectively acting as a void). In a widely reported experiment carried out by Capasso’s group at Harvard in 2009, a *repulsive* Casimir-Lifshitz force was observed between a silica plate and a polystyrene sphere coated with a gold film, immersed in a fluid of bromobenzene [44].

## References


1. H.B.G. Casimir, Koninkl. Ned. Akad. Wetenschap. **51**, 793 (1948)
2. S.K. Lamoreaux, Phys. Rev. Lett. **78** (1997)
3. M.J. Sparnaay, Physica **24**(6–10), 751–764 (1958)
4. P.H.G.M. van Blokland, J.T.G. Overbeek, J. Chem. Soc. Faraday Trans. **1**(74), 2637–2651 (1978)
5. A. Roy, C.-Y. Lin, U. Mohideen, Phys. Rev. D **60**, 111101 (1999)
6. J.D. Jackson, *Classical Electrodynamics* (Wiley, New York, 1998)
7. A. Zee, *Quantum Field Theory in a Nutshell* (Princeton University Press, Princeton, 2010)
8. D.A.R. Dalvit, P.W. Milonni, D. Roberts, F. Rosa, *Casimir Physics : Lecture Notes in Physics*, vol 834, chapter Introduction (Springer, Berlin, 2011)
9. M. Bordag, G.L. Klimchitskaya, U. Mohideen, V.M. Mostepanenko, *Advances in the Casimir Effect* (Oxford University Press, Oxford, 2009)

10. M. Bordag, U. Mohideen, V.M. Mostepanenko, Phys. Rep. **353**, 1–205 (2001)
11. R. Eisenschitz, F. London, Z. Physik **60**(7–8), 491–527 (1930)
12. F. London, Trans. Faraday Soc. **33**(8–26) (1937)
13. E.J.W. Verwey, J.G. Overbeek, *Theory of the Stability of Lyophobic* (Elsevier, Amsterdam, 1947)
14. H.B.G. Casimir, D. Polder, Phys. Rev. **73**(4) (1948)
15. J.E. Lennard-Jones, Proc. R. Soc. Lond. A **106**(738) (1924)
16. W. Heitler, F. London, Z. Phys. **44**(6–7), 455–472 (1927)
17. I.-C. Lin, A.P. Seitsonen, M.D. Coutinho-Neto, I. Tavernelli, U. Rothlisberger, J. Phys. Chem. B **113**(4), 1127–1131 (2009)
18. J.N. Israelachvili, Q. Rev. Biophys. **6**(341–387) (1973)
19. J.N. Israelachvili, *Intermolecular and Surface Forces* (Academic Press, London, 1991)
20. I. de Pater, J.J. Lissauer, *Planetary Sciences* (Cambridge University Press, Cambridge, 2010)
21. T. Poppe, J. Blum, T. Henning, ApJ **533**(454) (2000)
22. D.J. Scheeresa, C.M. Hartzella, P. Sáncheza, M. Swiftb, Icarus **533**(454–471) (2000)
23. B.V. Derjaguin, L.D. Landau, Zh. Eksp. Teor. Fiz **11**(12), 802 (1941)
24. A. Sambale, S.Y. Buhmann, D.-G. Welsch, Phys. Rev. A **75**(042109) (2007)
25. S.Y. Buhmann, *Dispersion forces I* (Springer, Heidelberg, 2013)
26. B. Brunetti, F. Pirani, F. Vecchiocattivi, E. Luzzatti, Chem. Phys. Lett. **55**(3) (1978)
27. C.H. Chen, P.E. Siska, Y.T. Lee, J. Chem. Phys. **59**(601) (1973)
28. L.W. Bruch, Surf. Sci. **125**(1), 194–217 (1983)
29. G. Binnig, C.F. Quate, Ch. Gerber, Phys. Rev. Lett. **56**, 930–933 (1986)
30. J.E. Lennard-Jones, Trans. Faraday Soc. **28**(333) (1932)
31. R. Kubo, Rep. Prog. Phys. **29**(1), 255 (1966)
32. A.D. McLachlan, Proc. R. Soc. Lond. A **274**(1356), 80–90 (1963)
33. G.S. Agarwal, Phys. Rev. A **11**, 230–242 (1975)
34. J.M. Wylie, J.E. Sipe, Phys. Rev. A **30**, 1185–1193 (1984)
35. A.D. McLachlan, Mol. Phys. **7**(4), 381 (1964)
36. D. Raskin, P. Kusch, Phys. Rev. **179**, 712–721 (1969)
37. I. Carusotto, L. Pitaevskii, S. Stringari, G. Modugno, M. Inguscio, Phys. Rev. Lett. **95**, 093202 (2005)
38. R.E. Grisenti, W. Schöllkopf, J.P. Toennies, G.C. Hegerfeldt, T. Köhler, Phys. Rev. Lett. **83**, 1755–1758 (1999)
39. J.H. de Boer, Trans. Faraday Soc. **32**, 10–37 (1936)
40. H.C. Hamaker, Physica **4**(10), 1058–1072 (1937)
41. E.M. Lifshitz, Zh. Eksp. Teor. Fiz. **29**, 94 (1955)
42. I.E. Dzyaloshinskii, E.M. Lifshitz, L.P. Pitaevskii, Adv. Phys. **10**, 165 (1961)
43. I.E. Dzyaloshinskii, L.P. Pitaevskii, van der Waals forces in an inhomogeneous dielectric. Sov. Phys. JETP **9**, 1282 (1959)
44. J.N. Munday, F. Capasso, V.A. Parsegian, Nature **457**, 170 (2009)
45. A.W. Rodriguez, F. Capasso, S.G. Johnson, Nat. Photonics **5**, 211 (2011)
46. S.K. Lamoreaux, in *Casimir Physics*, ed. by D. Dalvit, P. Milonni, D. Roberts, F. da Rosa. Lecture Notes in Physics, vol 834 (Springer, Berlin, Heidelberg, 2011), pp. 219–248
47. A. Roy, U. Mohideen, Phys. Rev. Lett. **82**, 4380–4383 (1999)
48. K.A. Milton, Am. J. Phys. **79**, 697 (2011)
49. R.L. Jaffe, Phys. Rev. D **72**(021301(R)) (2005)

# Chapter 2
# Macroscopic Quantum Electrodynamics

*Photon, photon, shining bright! Diffracting through the lab. at night. Can even God, with all his might, measure thy position right?*

The author

## 2.1 Field Quantisation in Vacuum

### *2.1.1 Quantising the Light Field*

We now learn from an early age that light is a form of radiation and a wave phenomenon that admits a continuum of frequencies. The visible light that the human eye perceives unaided is but a part of a larger electromagnetic spectrum that includes high frequency ultra-violet light at one end and low-frequency radiowaves at the other, involving an interplay of oscillating electric and magnetic fields that is well-described by Maxwell's equations. It is also an increasingly familiar thought, since the early 20th century, that light is in some sense particulate; it deposits its energy in discrete packets called photons, as postulated by Max Planck in his explanation of blackbody radiation, and hypostatised by Einstein in his treatement of the photo-electric effect. In fact, the reveberating dualism between light *qua* 'wave' and light *qua* 'atoms' has been traced back as far as the philosophers of classical Greece and the Hindu schools of ancient India.

In the quantum theory of light, first formulated by the British scientist Paul Dirac, something of a reconciliation of these seemedly incompatible conceptions is achieved—at least, at a formal level. Dirac proposed a quantisation of the electromagnetic field, as described by Maxwell's equations, involving an ensemble of harmonic oscillators with discrete energy levels. These 'quanta' are called photons. We will briefly propound a simple version of this *quantum electrodynamics* here.



W.M.R. Simpson, *Surprises in Theoretical Casimir Physics*, Springer Theses,
DOI 10.1007/978-3-319-09315-4_2

### 2.1.2 *Maxwell's Equations*

In the vacuum, the phenomenon of light is characterised by the electric field **E** and the magnetic induction **B**. The classical electrodynamic field obeys Maxwell's equations,

$$\nabla \cdot \mathbf{B} = 0, \quad \nabla \times \mathbf{E} = -\frac{\partial \mathbf{B}}{\partial t}, \tag{2.1.1}$$

$$\nabla \cdot \mathbf{E} = 0, \quad \nabla \times \mathbf{B} = \frac{1}{c^2}\frac{\partial \mathbf{E}}{\partial t}, \tag{2.1.2}$$

with the added constraint that the fields vanish at infinity. A useful representation of the fields involves the vector potential **A**,

$$\mathbf{E} = -\frac{\partial \mathbf{A}}{\partial t}, \quad \mathbf{B} = \nabla \times \mathbf{A}. \tag{2.1.3}$$

Expressed in this form, the first two of Maxwell's equations (2.1.1) are automatically satisfied. By imposing the Coulomb guage,

$$\nabla \cdot \mathbf{A} = 0 \tag{2.1.4}$$

the third of Maxwell's equations (2.1.2) is also satisfied. Using the only remaining non-trivial equation (2.1.2) we can derive the wave equation of light:

$$\nabla \times \nabla \times \mathbf{A} + \frac{1}{c^2}\frac{\partial^2 \mathbf{A}}{\partial t^2} = 0. \tag{2.1.5}$$

In quantum field theory the classical amplitudes must be replaced by quantum observables. The postulated connection between the classical and the quantum fields is elegantly simple, however: the classical fields are the ensemble averages of the quantum fields, e.g.

$$\mathbf{E} = \langle \psi | \hat{\mathbf{E}} | \psi \rangle, \tag{2.1.6}$$

where $|\psi\rangle$ is the state vector of the system, and $\hat{\mathbf{E}}$ is the 'operator-valued' quantum electric field. From the linearity of quantum mechanics and of Maxwell's equations, it follows that the quantum operators $\hat{\mathbf{E}}$ and $\hat{\mathbf{B}}$ also obey both Maxwell's equations (2.1.1, 2.1.2) and the electromagnetic wave equation (2.1.5). The specifically quantum-mechanical character of the field operators, however, resides in a fundamental commutation relation. It can be shown that [1]

$$[\hat{\mathbf{E}}(\mathbf{r}, t), \hat{\mathbf{E}}(\mathbf{r}', t)] = [\hat{\mathbf{B}}(\mathbf{r}, t), \hat{\mathbf{B}}(\mathbf{r}', t)] = \mathbf{0}, \tag{2.1.7}$$

$$[\hat{\mathbf{E}}(\mathbf{r}, t), \hat{\mathbf{B}}(\mathbf{r}', t)] = \frac{i\hbar}{\epsilon_0}\nabla \times \delta^T(\mathbf{r} - \mathbf{r}'), \tag{2.1.8}$$

where $\delta^T$ is the transversal delta function. Embedded in this result is a second assumption concerning the Hamiltonian of light, which governs the time evolution of the field and describes the total energy of the system: the *quantum* Hamiltonian has the same structure as the classical energy of the electromagnetic field [2]:

$$\hat{H} = \frac{1}{2}\int (\hat{\mathbf{E}}\cdot\hat{\mathbf{E}} + \hat{\mathbf{B}}\cdot\hat{\mathbf{B}})\,\mathrm{d}V, \tag{2.1.9}$$

where the volume integration is over all space.

### 2.1.3 The Quantum Light Mode

It is useful for our purposes to focus primarily on the vector potential $\mathbf{A}$, which determines the electromagnetic field (2.1.4), and to expand this quantum operator in terms of an appropriate set of light modes:

$$\hat{\mathbf{A}}(\mathbf{r},t) = \sum_k \left(\mathbf{A}_k(\mathbf{r},t)\hat{a}_k + \mathbf{A}_k^*(\mathbf{r},t)\hat{a}_k^\dagger\right). \tag{2.1.10}$$

In this formalism the modes, which are a set of classical waves $\{\mathbf{A}_k(\mathbf{r},t)\}$ obeying the laws of electromagnetism, are conjoined with quantum amplitude operators $\{\hat{a}_k, \hat{a}_k^\dagger\}$, which are associable with discrete excitations of the field. In the case of normal modes, the quantum amplitudes behave as creation and annihilation operators, satisfying Bose commutation relations

$$[\hat{a}_k, \hat{a}_{k'}^\dagger] = \delta_{kk'}, \quad [\hat{a}_k, \hat{a}_{k'}] = 0. \tag{2.1.11}$$

For convenience, we consider the special case of monochromatic modes[1] that oscillate at single frequencies,

$$\mathbf{A}_k(\mathbf{r},t) = \mathbf{A}_k(\mathbf{r})\exp(-i\omega_k t). \tag{2.1.12}$$

Under this representation, the Hamiltonian (2.1.9) can be rendered very simply:

$$\hat{H} = \sum_k \hbar\omega_k \left(\hat{a}_k^\dagger\hat{a}_k + \frac{1}{2}\right). \tag{2.1.13}$$

The total energy of the field is thus the sum of the energies of the modes, where each mode $k$ carries an energy of $\hbar\omega_k\,(n + 1/2)$, and $n$ is the number of photons present in that mode.

[1] Monochromatic modes are stationary modes that conserve energy. We therefore expect the Hamiltonian to be the sum of the Hamiltonians of the individual modes.

### 2.1.4 Zero-Point Energy

The form of the Hamiltonian in Eq. (2.1.13) is ubiquitous in quantum field theory; it embodies the expression for the energy of a simple quantum harmonic oscillator with discrete energy levels,

$$E = \hbar\omega\,(n + 1/2)\,, \quad n \in \mathbb{Z}, \tag{2.1.14}$$

for which even the unexcited state of the system ($n = 0$) has a non-zero energy of $\hbar\omega/2$. This is the so-called 'zero-point energy' associated with the ground-state of the field. We see that each mode of the field is like a distinct oscillator; each represents a degree of freedom. But we also see that there are an infinite number of them in the electromagnetic field. It follows that even in the ground state of each mode—*ergo*, the absence of any photons in the field—the minimal value of the energy is

$$E_0 = \sum_k \frac{\hbar\omega_k}{2} = \infty. \tag{2.1.15}$$

It seems obvious that infinities of this sort are unfortunate artifacts. It is not so obvious how to 'fix' quantum field theory to prevent it from producing them. Still, as we have seen in Chap. 1, the zero point energy leads to experimentally confirmed results. One of these results is the Casimir force (1.1.23).

### 2.1.5 External Boundaries on a Quantum Field

As it stands, however, Eq. (2.1.15) fails to be physically meaningful on at least two counts: it is both infinite and without empirical reference. There is no reason to think that infinite plane waves exist in nature. One way to register these constraints is through the imposition of 'external' boundary conditions, in which the field strength and its derivatives are fixed at certain locations. In our calculation in Chap. 1, we required that the transversal components of the electric field and the normal component of the magnetic field should vanish at the locations of the mirrors. The frequencies $\omega_k$, in this case, refer to a set of standing waves associated with the cavity, and with an energy that varies with the size of the cavity. This energy is finite, consequent upon an appropriate regularisation in which we effectively acknowledge that such a cavity cannot support an infinite number of frequencies.[2]

However, this method of representing the presence of material bodies within the system is acutely limited: real materials are dispersive, responding differently to different frequencies; they are dissipative, transforming electromagnetic energy into currents; and their presence is felt over spatially extended regions. Moreover, electromagnetic fields are present within material bodies, as well as the vacuum, where

[2] See Sect. 1.1.2.

Casimir forces are also to be expected. To facilitate a more sophisicated approach to the interaction of light with materials, we must learn to quantise the electromagnetic field in media.

## 2.2 Field Quantisation in Media

Our concern is with material bodies interacting with each other through the quantum electrodynamic field. Since such bodies involve large numbers of bound, charged particles, a microscopic description of their interaction is impractical, and we must instead consider the effective influence of these particles on the electromagnetic field. Contrary to conventional wisdom, a canonical quantisation of the electromagnetic field in the presence of media can in fact be performed [3, 4]. However, the process is somewhat involved, and the problems we wish to discuss formally do not depend upon the details. We will content ourselves, then, with a more phenomenological approach,[3] for the purpose of recovering the critical results we require with minimal effort and familiarising the reader with some essential ideas.

### *2.2.1 The Macroscopic Maxwell Equations*

In differential form, using *SI* units, the classical electrodynamic field obeys the macroscopic Maxwell equations

$$\nabla \cdot \mathbf{B} = 0, \quad \nabla \times \mathbf{E} = -\frac{\partial \mathbf{B}}{\partial t}, \tag{2.2.1}$$

$$\nabla \cdot \mathbf{D} = 0, \quad \nabla \times \mathbf{H} = \frac{1}{c^2}\frac{\partial \mathbf{D}}{\partial t}, \tag{2.2.2}$$

where the fields $\mathbf{D}$ and $\mathbf{H}$ are effective electric and magnetic fields respectively, associated with the polarisation and magnetisation of the media. We do not concern ourselves with external free charges, in the Casimir Effect. However, in extending quantum electrodynamics from light in empty space to light in media, we must lay down a set of constitutive equations coupling the fields $\hat{\mathbf{E}}$ to $\hat{\mathbf{D}}$ and $\hat{\mathbf{H}}$ to $\hat{\mathbf{B}}$. In the linear response regime

$$\mathbf{D} = \epsilon_0 \epsilon \mathbf{E}, \quad \mathbf{B} = \mu_0 \mu \mathbf{H}, \quad \epsilon_0 \mu_0 = 1/c^2, \tag{2.2.3}$$

[3] The basic approach adopted in this section is developed in detail in [8–10]. However, we will not use the Lorentz force expression, which does not recover the standard results for forces in media, but an analogue of the Minkowski stress tensor. Pitaevskii's comments here are relevant [13, 14].

where $(\epsilon_0)\ \epsilon$ is the *permittivity* of (free) space and $(\mu_0)\ \mu$ is the *permeability* of (free) space. For inhomogeneous media, $\epsilon$ and $\mu$ will vary in space. In the quantisation of the light field, all classical fields must be replaced by operator-valued quantum observables. It is convenient to introduce frequency components of the form

$$\hat{\mathbf{f}} = \int_0^\infty \mathrm{d}\omega\, \hat{\mathbf{f}}(\omega) + \mathrm{H.c}, \tag{2.2.4}$$

for some quantum operator $\hat{\mathbf{f}}$, where H.c. denotes the Hermitian conjugate of the preceding term. The quantised Maxwell equations can now be written:

$$\nabla \cdot \hat{\mathbf{B}} = 0, \quad \nabla \times \hat{\mathbf{E}} - i\omega\hat{\mathbf{B}} = \mathbf{0}. \tag{2.2.5}$$

$$\nabla \cdot \hat{\mathbf{D}} = 0, \quad \nabla \times \hat{\mathbf{H}} + i\omega\hat{\mathbf{D}} = \mathbf{0}. \tag{2.2.6}$$

### 2.2.2 Quantum Noise

The Heisenberg uncertainty principle entails the presence of quantum fluctuations that must be felt by the properties of the medium. At this level of description, without invoking the additional apparatus involved in the canonical theory of macro-QED [4], we must follow Lifshitz' somewhat ad hoc procedure [5] of introducing noise polarisation and magnetisation terms into the descriptions of the fields:

$$\hat{\mathbf{D}} = \epsilon_0\epsilon\hat{\mathbf{E}} + \hat{\mathbf{P}}_N, \tag{2.2.7}$$

$$\hat{\mathbf{H}} = \frac{1}{\mu_0\mu}\hat{\mathbf{B}} - \hat{\mathbf{M}}_N. \tag{2.2.8}$$

Situated in the effective fields associated with the media, these sources generate internal noise currents and noise charge densities within the material:

$$\hat{\mathbf{j}}_N = -i\omega\hat{\mathbf{P}}_N + \nabla \times \hat{\mathbf{M}}_N, \tag{2.2.9}$$

$$\hat{\rho}_N = -\nabla \cdot \hat{\mathbf{P}}_N. \tag{2.2.10}$$

These source fields satisfy the continuity equation

$$-i\omega\hat{\rho}_N + \nabla \cdot \hat{\mathbf{j}}_N = 0. \tag{2.2.11}$$

In Lifshitz theory, the electromagnetic fields remain formally unquantised and classical, with the behavior of the noise fields being governed by the results of Rytov theory [6] to produce the appropriate stochastic behaviour [7]. Here, we will follow the more conspicuously quantum-mechanical approach developed in [8–10].

### 2.2.3 Bosonic Field Operators

Cast in the form of quantum operators, our noise fields must be made to satisfy quantum commutation relations. We choose to adopt relations such that the fluctuation spectrum obeys the fluctuation-dissipation theorem [11], entailing (among other things) that the noise should vanish *on average*. This can be achieved by appropriately relating the polarisation and magnetisation to bosonic creation and annihilation operators $\hat{\mathbf{f}}_\lambda^\dagger(\mathbf{r},\omega)$ and $\hat{\mathbf{f}}_\lambda(\mathbf{r},\omega)$ for the electric and magnetic contributions to the field $\lambda \in \{e, m\}$,

$$\hat{\mathbf{P}}_N(\mathbf{r},\omega) = i\sqrt{\frac{\hbar\epsilon_0}{\pi}\,\mathrm{Im}\,\epsilon(\mathbf{r},\omega)}\,\hat{\mathbf{f}}_e(\mathbf{r},\omega), \tag{2.2.12}$$

$$\hat{\mathbf{M}}_N(\mathbf{r},\omega) = i\sqrt{\frac{\hbar}{\pi\mu_0}\frac{\mathrm{Im}\,\mu(\mathbf{r},\omega)}{|\mu(\mathbf{r},\omega)|^2}}\,\hat{\mathbf{f}}_m(\mathbf{r},\omega), \tag{2.2.13}$$

which themselves obey bosonic commutation relations:

$$\left[\hat{\mathbf{f}}_\lambda(\mathbf{r},\omega), \hat{\mathbf{f}}_{\lambda'}(\mathbf{r}',\omega')\right] = \left[\hat{\mathbf{f}}_\lambda^\dagger(\mathbf{r},\omega), \hat{\mathbf{f}}_{\lambda'}^\dagger(\mathbf{r}',\omega')\right] = \mathbf{0}, \tag{2.2.14}$$

$$\left[\hat{\mathbf{f}}_\lambda(\mathbf{r},\omega), \hat{\mathbf{f}}_{\lambda'}^\dagger(\mathbf{r}',\omega')\right] = \delta_{\lambda\lambda'}\delta(\mathbf{r}-\mathbf{r}')\delta(\omega-\omega'). \tag{2.2.15}$$

In fact, these operators will serve to describe the collective, polariton-like, bosonic excitations of the body-field system, for which we may define a system ground-state:

$$\hat{\mathbf{f}}_\lambda(\mathbf{r},\omega)\,|\{0\}\rangle = \mathbf{0} \quad \forall\lambda,\mathbf{r},\omega. \tag{2.2.16}$$

When the system is in its ground state, the electromagnetic field is likewise ('the quantum vacuum'). A complete Hilbert-space spanned by Fock states can be obtained in the usual way by repeated application of creation operators $\hat{\mathbf{f}}_\lambda^\dagger(\mathbf{r},\omega)$ to the ground state:

$$\left|\{n_{\lambda_1}(\mathbf{r}_1,\omega_1), n_{\lambda_2}(\mathbf{r}_2,\omega_2), \ldots,\}\right\rangle = \prod_{k,n_{\lambda_k}} \frac{1}{\sqrt{n_{\lambda_k}(\mathbf{r}_k,\omega_k)!}}\hat{\mathbf{f}}_{\lambda_k}^\dagger(\mathbf{r}_k,\omega_k)\,|\{0\}\rangle\,. \tag{2.2.17}$$

### 2.2.4 Field Fluctuations

For a system prepared in a state $|\phi\rangle$, where $|\phi\rangle$ is represented as a vector of a Hilbert space, the quantum average of an observable $\hat{q}$ is given by

$$\langle\hat{q}\rangle = \langle\phi|\,\hat{q}\,|\phi\rangle\,. \tag{2.2.18}$$

The fluctuations associated with a quantum operator are

$$\left\langle (\Delta\hat{q})^2 \right\rangle = \left\langle \hat{q}^2 \right\rangle - \langle \hat{q} \rangle^2 . \tag{2.2.19}$$

As noted, these fluctuations necessarily occur for two non-commuting operators as a direct consequence of the Heisenberg uncertainty principle:

$$\left\langle \left(\Delta\hat{f}\right)^2 (\Delta\hat{g})^2 \right\rangle \geq \frac{1}{4} \left|\left\langle \left[\hat{f}, \hat{g}\right]\right\rangle\right|^2 . \tag{2.2.20}$$

Our field operators $\hat{f}$ clearly have a vanishing ground-state average:

$$\begin{aligned} \langle\{0\}| \, \hat{\mathbf{f}}_\lambda(\mathbf{r}, \omega)) \, |\{0\}\rangle &= \mathbf{0}, \\ \langle\{0\}| \, \hat{\mathbf{f}}_\lambda^\dagger(\mathbf{r}, \omega)) \, |\{0\}\rangle &= \mathbf{0}. \end{aligned} \tag{2.2.21}$$

It follows that the noise polarisation and magnetisation operators, defined in Eqs. (2.2.12) and (2.2.13), consequently have zero-averages as well:

$$\begin{aligned} \left\langle \hat{\mathbf{P}}_N(\mathbf{r}, \omega) \right\rangle &= 0, \\ \left\langle \hat{\mathbf{M}}_N(\mathbf{r}, \omega) \right\rangle &= 0. \end{aligned} \tag{2.2.22}$$

For paired field operators, we obtain the following results:

$$\begin{aligned} \left\langle \hat{\mathbf{f}}_\lambda(\mathbf{r}, \omega)\hat{\mathbf{f}}_{\lambda'}(\mathbf{r}', \omega') \right\rangle &= \mathbf{0}, \\ \left\langle \hat{\mathbf{f}}_\lambda(\mathbf{r}, \omega)\hat{\mathbf{f}}^\dagger_{\lambda'}(\mathbf{r}', \omega') \right\rangle &= \delta_{\lambda\lambda'}\delta(\mathbf{r} - \mathbf{r}')\delta(\omega - \omega'), \\ \left\langle \hat{\mathbf{f}}^\dagger_\lambda(\mathbf{r}, \omega)\hat{\mathbf{f}}_{\lambda'}(\mathbf{r}', \omega') \right\rangle &= \mathbf{0}, \\ \left\langle \hat{\mathbf{f}}^\dagger_\lambda(\mathbf{r}, \omega)\hat{\mathbf{f}}^\dagger_{\lambda'}(\mathbf{r}', \omega') \right\rangle &= \mathbf{0}. \end{aligned} \tag{2.2.23}$$

From these, it follows that the noise polarisation and magnetisation have non-zero fluctuations, leading to the expressions [10]

$$\left\langle S\left[\Delta\hat{\mathbf{P}}_N(\mathbf{r}, \omega)\Delta\hat{\mathbf{P}}^\dagger_N(\mathbf{r}', \omega')\right]\right\rangle = \frac{\hbar}{2\pi}\epsilon_0 \mathrm{Im}\, \chi(\mathbf{r}, \omega)\delta(\mathbf{r} - \mathbf{r}')\delta(\omega - \omega'), \tag{2.2.24}$$

$$\left\langle S\left[\Delta\hat{\mathbf{M}}_N(\mathbf{r}, \omega)\Delta\hat{\mathbf{M}}^\dagger_N(\mathbf{r}', \omega')\right]\right\rangle = \frac{\hbar}{2\pi}\frac{\mathrm{Im}\, \zeta(\mathbf{r}, \omega)}{\mu_0}\delta(\mathbf{r} - \mathbf{r}')\delta(\omega - \omega'), \tag{2.2.25}$$

where the electric susceptibility $\chi(\mathbf{r}, \omega)$ is related to the electric permittivity via

$$\epsilon(\mathbf{r},\omega) = 1 + \chi(\mathbf{r},\omega), \tag{2.2.26}$$

the magnetic susceptibility $\zeta(\mathbf{r},\omega)$ is related to the magnetic permeability via

$$\mu(\mathbf{r},\omega) = \frac{1}{1-\zeta(\mathbf{r},\omega)}, \tag{2.2.27}$$

and $S$ denotes a symmetrised operator product

$$S\left[\hat{a}\hat{b}\right] = \frac{1}{2}\left(\hat{a}\hat{b} + \hat{b}\hat{a}\right). \tag{2.2.28}$$

This result concurs with the fluctuation-dissipation theorem [11], which relates the fluctuations of a quantity with the rate of absorption of energy by the system when an external force is applied. In fact, we intentionally defined (2.2.12) and (2.2.13) in order for them to do so. This is the same spectrum recovered by Rytov theory, and its application to the electromagnetic field lies at the heart of the Lifshitz calculation of the Casimir force.

### 2.2.5 The Fundamental Fields

It is now possible to express the other fields in terms of the fundamental field operators. Of course, we achieve this by solving Maxwell's equations. Noting that the noise operators we have introduced have effectively created internal currents (2.2.9) and charges (2.2.10) in the system, we can use the inhomogeneous Helmholtz equation for the electric field

$$\left[\nabla\times\frac{1}{\mu}\nabla\times-\frac{\omega^2}{c^2}\epsilon\right]\hat{\mathbf{E}} = i\mu_0\omega\hat{\mathbf{j}}_N. \tag{2.2.29}$$

This equation is formally solveable by means of a classical Green tensor characterising the linear response of the field to the current sources:

$$\hat{\mathbf{E}}(\mathbf{r},\omega) = i\mu_0\omega\int \mathrm{d}^3r'\,\mathbf{G}(\mathbf{r},\mathbf{r}',\omega)\cdot\hat{\mathbf{j}}_N(\mathbf{r}',\omega). \tag{2.2.30}$$

The noise current (2.2.9) was earlier expressed in terms of the noise polarisation (2.2.12) and magnetisation (2.2.13), and these have been rewritten in terms of the fundamental field operators $\hat{\mathbf{f}}_e(\mathbf{r},\omega)$ and $\hat{\mathbf{f}}_m(\mathbf{r},\omega)$. With reference to Eqs. (2.2.9), (2.2.12) and (2.2.13), an expression for the field can be straightforwardly written as

$$\hat{\mathbf{E}}(\mathbf{r}) = \int_0^\infty \mathrm{d}\omega\sum_{\lambda=e,m}\int \mathrm{d}^3r'\,\mathbf{G}_\lambda(\mathbf{r},\mathbf{r}',\omega)\cdot\hat{f}_\lambda(\mathbf{r}',\omega) + \text{H.c}, \tag{2.2.31}$$

where

$$\mathbf{G}_e(\mathbf{r}, \mathbf{r}', \omega) = i\frac{\omega^2}{c^2}\left(\frac{\hbar}{\pi\epsilon_0}\,\mathrm{Im}\,\epsilon(\mathbf{r}', \omega)\right)^{1/2} \mathbf{G}(\mathbf{r}, \mathbf{r}', \omega), \tag{2.2.32}$$

$$\mathbf{G}_m(\mathbf{r}, \mathbf{r}', \omega) = i\frac{\omega}{c}\left(\frac{\hbar}{\pi\epsilon_0}\frac{\mathrm{Im}\,\mu(\mathbf{r}', \omega)}{|\mu(\mathbf{r}, \omega)|^2}\right)^{1/2} \left[\nabla \times \mathbf{G}(\mathbf{r}, \mathbf{r}', \omega)\right]^T. \tag{2.2.33}$$

The other fields may also be expressed similarly. Using Eq. (2.2.31) and the Maxwell equation (2.2.5), we obtain

$$\hat{\mathbf{B}}(\mathbf{r}) = \int_0^\infty \frac{\mathrm{d}\omega}{i\omega} \sum_{\lambda=e,m} \int \mathrm{d}^3 r'\, \nabla \times \mathbf{G}_\lambda(\mathbf{r}, \mathbf{r}', \omega) \cdot \hat{f}_\lambda(\mathbf{r}', \omega) + \mathrm{H.c.} \tag{2.2.34}$$

From (2.2.7) and (2.2.12), conjoined with (2.2.31) we obtain

$$\hat{\mathbf{D}}(\mathbf{r}) = \int_0^\infty \mathrm{d}\omega \left[\epsilon_0\epsilon(\mathbf{r}, \omega) \sum_{\lambda=e,m} \int \mathrm{d}^3 r' \mathbf{G}_\lambda(\mathbf{r}, \mathbf{r}') \cdot \mathbf{f}_\lambda(\mathbf{r}', \omega) \right.$$
$$\left. + i\left(\frac{\hbar\epsilon_0}{\pi}\mathrm{Im}\,\epsilon(\mathbf{r}, \omega)\right)^{1/2} \mathbf{f}_e(\mathbf{r}, \omega)\right] + \mathrm{H.c.} \tag{2.2.35}$$

And from (2.2.8) and (2.2.13), conjoined with (2.2.34) we obtain

$$\mathbf{H}(\mathbf{r}) = \int_0^\infty \mathrm{d}\omega \left[\frac{1}{i\omega\mu_0\mu(\mathbf{r}, \omega)} \sum_{e,m} \int \mathrm{d}^3 r' \nabla \times \mathbf{G}_\lambda(\mathbf{r}, \mathbf{r}\,\omega) \cdot \mathbf{f}_\lambda(\mathbf{r}', \omega) \right.$$
$$\left. - \left(\frac{\hbar\kappa_0}{\pi}\frac{\mathrm{Im}\,\mu(\mathbf{r}, \omega)}{\mu(\mathbf{r}, \omega)}\right)^{1/2} \mathbf{f}_m(\mathbf{r}, \omega)\right] + \mathrm{H.c.} \tag{2.2.36}$$

With a little further work [8], we can recover the commutation relations

$$\left[\mathbf{E}(\mathbf{r}), \mathbf{E}(\mathbf{r}')\right] = \left[\mathbf{B}(\mathbf{r}), \mathbf{B}(\mathbf{r}')\right] = \mathbf{0}, \tag{2.2.37}$$

$$\left[\mathbf{E}(\mathbf{r}), \mathbf{B}(\mathbf{r}')\right] = \frac{i\hbar}{\epsilon_0}\nabla \times \delta(\mathbf{r} - \mathbf{r}'), \tag{2.2.38}$$

which agree with those in free-space (2.1.8), and deduce the ground-state fluctuation spectrum of the field [10]

$$\left\langle S\left[\Delta\hat{\mathbf{E}}(\mathbf{r}, \omega)\Delta\hat{\mathbf{E}}^\dagger(\mathbf{r}', \omega')\right]\right\rangle = \frac{\hbar}{2\pi}\omega^2\mu_0 \mathrm{Im}\,\mathbf{G}(\mathbf{r}, \mathbf{r}'\,\omega)\delta(\omega - \omega'), \tag{2.2.39}$$

which is in accordance with the fluctuation-dissipation theorem [11].

### 2.2.6 The Hamiltonian

The Hamiltonian governs the dynamics of a system. Since the electromagnetic field operators we introduced are linear combinations of the fundamental field operators, our Hamiltonian must therefore generate the correct time-dependence of the field operators, so that Maxwell's equations and the constitutive equations hold. This behaviour is implemented using the Hamiltonian

$$\hat{H} = \sum_{\lambda=e,m} \int \mathrm{d}^3r \int_0^\infty \mathrm{d}\omega\, \hbar\omega\, \hat{\mathbf{f}}_\lambda^\dagger(\mathbf{r},\omega)\cdot \hat{\mathbf{f}}_\lambda(\mathbf{r},\omega). \tag{2.2.40}$$

From the Heisenberg equation of motion, we find

$$\frac{\mathrm{d}}{\mathrm{d}t}\hat{\mathbf{f}}_\lambda = \frac{1}{i\hbar}\left[\hat{\mathbf{f}}_\lambda(\mathbf{r},\omega),\hat{H}\right] = -i\omega\hat{\mathbf{f}}_\lambda(\mathbf{r},\omega). \tag{2.2.41}$$

This differential equation is solved by

$$\hat{\mathbf{f}}_\lambda(\mathbf{r},\omega,t) = \hat{\mathbf{f}}_\lambda(\mathbf{r},\omega)\mathrm{e}^{-i\omega t}. \tag{2.2.42}$$

This gives the correct behaviour for recovering the Maxwell equations: the time-dependent frequency components of the electromagnetic fields are ordinary Fourier components. The ground state defined earlier is clearly an eigenstate of this Hamiltonian,

$$\hat{H}\,|\{0\}\rangle = 0. \tag{2.2.43}$$

Using the bosonic commutation relations (2.2.15), we find that

$$\hat{H}\,|n_\lambda(\mathbf{r},\omega)\rangle = \hbar\omega\, n_\lambda(\mathbf{r},\omega)\,|n_\lambda(\mathbf{r},\omega)\rangle\,, \tag{2.2.44}$$

and more generally that

$$\begin{aligned}&\hat{H}\left|n_{\lambda_1}(\mathbf{r}_1,\omega_1),\, n_{\lambda_2}(\mathbf{r}_2,\omega_2),\ldots\right\rangle\\ &\quad = \hbar\left(n_{\lambda_1}\omega_1 + n_{\lambda_2}\omega_2 + \ldots\right)\left|n_{\lambda_1}(\mathbf{r}_1,\omega_1),\, n_{\lambda_2}(\mathbf{r}_2,\omega_2),\ldots\right\rangle,\end{aligned} \tag{2.2.45}$$

so a multi-mode quantum Fock state is an energy eigenstate, where the energy is the sum of the energies associated with each individual excitation.

### 2.2.7 Photons and Polaritons

Significantly, the quantum of the interacting system of dielectric material and electromagnetic fields is no longer that of the photon; the Hamiltonian has been cast in the form of a summation over field operators in which the material and electromagnetic components of the system have been *mixed*. This is a kind of polariton, and the quantisation of the coupled system in terms of polaritons has implications for how we understand the *nature* of the Casimir force. This question will be explored further in Chap. 3. At this point, we will content ourselves simply with observing that the Hamiltonian of (2.1.9) can be related to the Hamiltonian of (2.2.40) as the limiting case in which the refractive index of the material (or its resistance to the field, if you will) is so high that only a negligible fraction of the radiation can be absorbed. In such idealised cases, the polariton is almost entirely 'photonic'.

## 2.3 The Casimir Force Density

### 2.3.1 The Stress Tensor

As we have seen, the ground-state of the coupled system of electromagnetic field and dielectric is one with non-zero current density within the medium, consistent with the fluctuation-dissipation theorem. The Casimir force arises from the interaction of these currents. In Lifshitz' original theory, the force was obtained by averaging Maxwell's stress tensor in a vacuum with respect to electromagnetic fluctuations [5],

$$\sigma_F = (\mathbf{E} \otimes \mathbf{E}) + (\mathbf{B} \otimes \mathbf{B}) - \frac{1}{2}(\mathbf{E} \cdot \mathbf{E} + \mathbf{B} \cdot \mathbf{B})\,\mathbf{1}_3, \tag{2.3.1}$$

which governs the flow of momentum associated with the electromagnetic field.[4] In the extension of Lifshitz theory to the general case of Casimir forces between bodies embedded in media [12], the Casimir (or 'Casimir-Lifshitz') forces in a system are ultimately determined by an analogue of the Minkowski stress tensor

$$\sigma_M = (\mathbf{D} \otimes \mathbf{E}) + (\mathbf{B} \otimes \mathbf{H}) - \frac{1}{2}(\mathbf{D} \cdot \mathbf{E} + \mathbf{B} \cdot \mathbf{H})\,\mathbf{1}_3, \tag{2.3.2}$$

which concerns the momentum associated with the body-assisted field, reducing to the Maxwell stress tensor in the vacuum. Here we will derive a general expression for the Casimir stress tensor in an inhomogeneous medium, relating the Minkoswki stress tensor to the Casimir force density. The derived expression also applies to piece-wise homogeneous systems and an infinite homogeneous medium as special cases. The subject of the Casimir stress tensor has been a matter of some controversy.[5] However, we offer here a fresh and simple argument that coincides with the results

---

[4] See Appendix B and [21].

[5] The accepted use of the 'Minkowski-like' stress tensor for computing Casimir forces in the Lifshitz theory [22] was challenged in [23], resulting in some debate [13, 24, 25]. A new argument for the

of Lifshitz theory[6]; the standard predictions may be recovered after quantising the designated stress tensor at the end of what is essentially a classical demonstration.

It must be emphasised from the outset that this is intended more as an illustrative argument than a 'water-tight' proof. Once the effects of temperature are considered and absorption is included, the derivation becomes substantially more involved, but the final result for the stress tensor is the same [3, 12].

#### 2.3.1.1 Prior Conditions on the Stress Tensor

The stress tensor describing the momentum flow for an object embedded in a fluid medium experiencing the fluctuations of the electromagnetic field must satisfy certain properties, as discussed at length by Pitaevskii [13, 14]. It must incorporate both the electromagnetic and fluid-mechanical aspects of the problem, and be decomposable into the form

$$\sigma = -P_0(\rho)\mathbb{1}_3 + \sigma_E, \tag{2.3.3}$$

where $P_0$ is defined as the pressure of a uniform infinite liquid of density $\rho$ in the absence of electromagnetic fluctuations, and $\sigma_E$ is the contribution to the stress in the system arising specifically from the fluctuations of the field. For our purposes, temperature is not a variable; whilst the purely quantum-mechanical Casimir force requires thermodynamic equilibrium, it occurs at zero temperature. In addition, we require that the stress tensor must be symmetric: $\sigma = \sigma^T$.

The physical picture is this: electromagnetic fluctuations in the fluid exert a radiation pressure on the molecules that compose it. Mechanical equilibrium is ensured, however, by the presence of a counteracting pressure term in the stress tensor (2.3.3) preventing a permanent flow in the liquid (which would lead to a *perpetuum mobile*). The total stress tensor must therefore be derived in the circumstances of both thermodynamic and mechanical equilibrium, and consequently we require that

$$\nabla \cdot \sigma = 0. \tag{2.3.4}$$

#### 2.3.1.2 The Pressure Force

We consider a small deformation at the surface of a body with a displacement vector field $\delta\mathbf{r}(\mathbf{r})$. The change in the free energy is

$$\delta F = -\int \mathrm{d}V\, \mathbf{f} \cdot \delta\mathbf{r}, \tag{2.3.5}$$

(Footnote 5 continued)
disputed result can be found in [3], where the Casimir stress was derived in the context of the canonical theory of macroscopic quantum electrodynamics [4].

[6] Lifshitz theory, in this case, refers to the more general results obtained in [12].

where $\mathbf{f}$ is the force per unit volume on the body during the deformation. From the continuity equation of fluid dynamics, we infer that

$$\delta\rho + \nabla \cdot (\rho\delta\mathbf{r}) = 0 \Longrightarrow \delta\rho = -\nabla \cdot (\rho\,\delta\mathbf{r})\,. \tag{2.3.6}$$

The variation of the permittivity $\epsilon$ is related to the material deformation via

$$\delta\epsilon = \frac{\partial\epsilon}{\partial\rho}\delta\rho = -\frac{\partial\epsilon}{\partial\rho}\nabla \cdot (\rho\,\delta\mathbf{r})\,. \tag{2.3.7}$$

Expanding the divergence, and observing that

$$\frac{\partial\epsilon}{\partial\rho}\delta\mathbf{r} \cdot \nabla\rho = \nabla\epsilon \cdot \delta\mathbf{r}, \tag{2.3.8}$$

we obtain

$$\delta\epsilon = -\delta\mathbf{r} \cdot \nabla\epsilon - \rho\frac{\partial\epsilon}{\partial\rho}\nabla \cdot \delta\mathbf{r}, \tag{2.3.9}$$

where $\epsilon = \epsilon\,(\rho)$ and $\rho = \rho\,(\mathbf{r})$. The variation of the permeability $\mu$ is governed similarly. The energy associated with the electromagnetic field in media is of the form [2]

$$E_f = \frac{1}{2}\int \left(\epsilon\,E^2 + \frac{1}{\mu}B^2\right) \mathrm{d}V. \tag{2.3.10}$$

We postulate that the variation in the free energy[7] is of the form [15]

$$\delta F = \delta F_0 - \frac{1}{2}\int \left(\delta\epsilon\,E^2 + \delta\left(\mu^{-1}\right) B^2\right) \mathrm{d}V. \tag{2.3.11}$$

Inserting (2.3.9) and its analogue for $\mu$ into (2.3.11), we obtain

$$\begin{aligned}\delta F = \delta F_0 &+ \frac{1}{2}\int \left[(\nabla\epsilon)\,E^2 + \left(\nabla\frac{1}{\mu}\right) B^2\right] \cdot \delta\mathbf{r}\,\mathrm{d}V \\ &+ \frac{1}{2}\int \left(E^2\rho\frac{\partial\epsilon}{\partial\rho}\nabla \cdot \delta\mathbf{r} + B^2\rho\frac{\partial}{\partial\rho}\left(\frac{1}{\mu}\right)\nabla \cdot \delta\mathbf{r}\right) \mathrm{d}V.\end{aligned} \tag{2.3.12}$$

Using integration by parts, we find that

---

[7] This result is not valid when there is absorption. Note that the expression for the variation of the free energy in Lifshitz theory takes a similar form [12]:

$$\delta F = \delta F_0 - \frac{T}{4\pi}\sum_{n=0}^{\infty}\int D_{ii}(\mathbf{r}, \mathbf{r}, \xi_n)\delta\epsilon(\mathbf{r}, i\xi_n)\mathrm{d}^3\mathbf{r}.$$

$$\int \left[ E^2 \left( \rho \frac{\partial \epsilon}{\partial \rho} \nabla \cdot \delta \mathbf{r} \right) \right] \mathrm{d}V = \int_{\partial V} E^2 \left( \rho \frac{\partial \epsilon}{\partial \rho} \delta \mathbf{r} \right) \mathrm{d}A - \int \nabla \left[ E^2 \left( \rho \frac{\partial \epsilon}{\partial \rho} \right) \right] \cdot \delta \mathbf{r} \, \mathrm{d}V. \tag{2.3.13}$$

For a sufficiently large volume, the field terms vanish on the surface. We deduce that

$$\int \left[ E^2 \left( \rho \frac{\partial \epsilon}{\partial \rho} \nabla \cdot \delta \mathbf{r} \right) \right] \mathrm{d}V = - \int \nabla \left[ E^2 \left( \rho \frac{\partial \epsilon}{\partial \rho} \right) \right] \cdot \delta \mathbf{r} \, \mathrm{d}V. \tag{2.3.14}$$

The terms in $\mu$ may be treated similarly. Hence Eq. (2.3.12) may be recast in the form

$$\delta F = \delta F_0 + \frac{1}{2} \int \mathrm{d}V \left\{ (\nabla \epsilon) E^2 + \left( \nabla \frac{1}{\mu} \right) B^2 - \nabla \left[ E^2 \left( \rho \frac{\partial \epsilon}{\partial \rho} \right) \right] - \nabla \left[ B^2 \left( \rho \frac{\partial}{\partial \rho} \frac{1}{\mu} \right) \right] \right\} \cdot \delta \mathbf{r}. \tag{2.3.15}$$

The free energy is clearly associated, via Eq. (2.3.5), with a force of the form

$$\mathbf{f} = -\nabla P_0 - \frac{1}{2} (\nabla \epsilon) E^2 - \frac{1}{2} \left( \nabla \frac{1}{\mu} \right) B^2 + \frac{1}{2} \nabla \left[ E^2 \left( \rho \frac{\partial \epsilon}{\partial \rho} \right) + B^2 \left( \rho \frac{\partial}{\partial \rho} \frac{1}{\mu} \right) \right], \tag{2.3.16}$$

acting per unit volume of the body. We consider the case of mechanical equilibrium, in which the forces on the body are balanced, i.e.

$$\mathbf{f} = 0 \implies \mathbf{f}_P + \mathbf{f}_{\mathcal{E}} = \mathbf{0}, \tag{2.3.17}$$

where we decompose the force into two contributions:

$$\begin{aligned} \mathbf{f}_P &= -\nabla P_0 + \frac{1}{2} \nabla \left[ E^2 \left( \rho \frac{\partial \epsilon}{\partial \rho} \right) + B^2 \left( \rho \frac{\partial}{\partial \rho} \frac{1}{\mu} \right) \right], \\ \mathbf{f}_{\mathcal{E}} &= -\frac{1}{2} \left[ (\nabla \epsilon) E^2 + \nabla \left( \mu^{-1} \right) B^2 \right]. \end{aligned} \tag{2.3.18}$$

It will become clear that these balancing contributions can be assigned distinct physical meanings.[8] Our expression for the total force (2.3.17) may be recast in the form of a stress tensor $\sigma$ satisfying

$$\mathbf{f} = \nabla \cdot \sigma = 0. \tag{2.3.19}$$

[8] We note that the first contribution $\mathbf{f}_P$ contains a pressure term which is present in the absence of electromagnetic fluctuations, and a contribution due to the deformation of the medium which vanishes in the limit of an incompressible medium.

It follows that the complete stress tensor takes the form

$$\sigma = -P_0 \mathbf{1}_3 + \frac{1}{2}\left\{E^2\left(\rho\frac{\partial\epsilon}{\partial\rho}\right) + B^2\left(\rho\frac{\partial}{\partial\rho}\frac{1}{\mu}\right)\right\}\mathbf{1}_3 \\ -\frac{1}{2}\left\{\int \nabla\epsilon(\mathbf{r})E^2(\mathbf{r})\cdot d\mathbf{r} + \int \nabla\left(\frac{1}{\mu(\mathbf{r})}\right)B^2(\mathbf{r})\cdot d\mathbf{r}\right\}\mathbf{1}_3, \tag{2.3.20}$$

which we similarly decompose into two contributions

$$\sigma = \sigma_P + \sigma_{\mathcal{E}}, \tag{2.3.21}$$

where

$$\sigma_P = -P_0 \mathbf{1}_3 + \frac{1}{2}\left\{E^2\left(\rho\frac{\partial\epsilon}{\partial\rho}\right) + B^2\left(\rho\frac{\partial}{\partial\rho}\frac{1}{\mu}\right)\right\}\mathbf{1}_3, \tag{2.3.22}$$

$$\sigma_{\mathcal{E}} = -\frac{1}{2}\left\{\int \nabla\epsilon(\mathbf{r})E^2(\mathbf{r})\cdot d\mathbf{r} + \int \nabla\left(\frac{1}{\mu(\mathbf{r})}\right)B^2(\mathbf{r})\cdot d\mathbf{r}\right\}\mathbf{1}_3. \tag{2.3.23}$$

The two force densities may be defined with respect to the stress components:

$$\mathbf{f}_P = \nabla\cdot\sigma_P, \quad \mathbf{f}_{\mathcal{E}} = \nabla\cdot\sigma_{\mathcal{E}}. \tag{2.3.24}$$

#### 2.3.1.3 The Minkowski Contribution

The term we have labelled $\sigma_{\mathcal{E}}$ can be related to the Minkowski stress tensor via Maxwell's equations, in the absence of external charges and under the condition of equilibrium. We begin with the general form of the classical Maxwell equations:

$$\nabla\cdot\mathbf{B} = 0, \quad \nabla\times\mathbf{E} = -\frac{\partial\mathbf{B}}{\partial t}, \quad \nabla\cdot\mathbf{D} = \rho, \quad \nabla\times\mathbf{H} = \frac{\partial\mathbf{D}}{\partial t} + \mathbf{J}. \tag{2.3.25}$$

From the first equation, it follows that

$$\nabla(\mathbf{B}\otimes\mathbf{H}) = \mathbf{H}(\nabla\cdot\mathbf{B}) + (\mathbf{B}\cdot\nabla)\mathbf{H} = (\mathbf{B}\cdot\nabla)\mathbf{H}. \tag{2.3.26}$$

From the third, we deduce

$$\nabla(\mathbf{D}\otimes\mathbf{E}) = \mathbf{E}(\nabla\cdot\mathbf{D}) + (\mathbf{D}\cdot\nabla)\mathbf{E} = \rho\mathbf{E} + (\mathbf{D}\cdot\nabla)\mathbf{E}. \tag{2.3.27}$$

Combining (2.3.26) and (2.3.27), we obtain

$$\nabla(\mathbf{D}\otimes\mathbf{E}) + \nabla(\mathbf{B}\otimes\mathbf{H}) = \rho\mathbf{E} + (\mathbf{D}\cdot\nabla)\mathbf{E} + (\mathbf{B}\cdot\nabla)\mathbf{H}. \tag{2.3.28}$$

From the second and fourth equations, we find

$$\frac{\partial}{\partial t}(\mathbf{D} \times \mathbf{B}) = (\nabla \times \mathbf{E}) \times \mathbf{D} + (\nabla \times \mathbf{H}) \times \mathbf{B} - \mathbf{J} \times \mathbf{B}. \qquad (2.3.29)$$

It follows from (2.3.28) and (2.3.29) that

$$\mathbf{F}_L + \frac{\partial}{\partial t}(\mathbf{D} \times \mathbf{B}) = \mathbf{T}, \qquad (2.3.30)$$

where $\mathbf{F}_L$ is the Lorentz force that acts on external charges and currents in the system,

$$\mathbf{F}_L = \rho \mathbf{E} + \mathbf{J} \times \mathbf{B}, \qquad (2.3.31)$$

and $\mathbf{T}$ can be decomposed into the sum of two terms:

$$\mathbf{T}_1 = (\nabla \times \mathbf{E}) \times \mathbf{D} - (\mathbf{D} \cdot \nabla)\,\mathbf{E} + (\nabla \times \mathbf{H}) \times \mathbf{B} - (\mathbf{B} \cdot \nabla)\,\mathbf{H}, \qquad (2.3.32)$$

$$\mathbf{T}_2 = \nabla\,(\mathbf{D} \otimes \mathbf{E}) + \nabla\,(\mathbf{B} \otimes \mathbf{H}). \qquad (2.3.33)$$

Both sides of Eq. (2.3.30) must have units of force density. However, in macroscopic electromagnetism applied solely to dielectrics (including metals) there are no external currents or charges: $\rho = 0$ and $\mathbf{J} = 0$. It follows that the Lorentz force is identically zero:

$$\mathbf{F}_L = 0. \qquad (2.3.34)$$

Whilst a microscopic Lorentz force could be defined with respect to the *internal* source terms introduced in the Lifshitz theory, these terms are not the referents of (2.3.25), and this approach does not recover the accepted expression for the forces in media [13, 14]. Also, we will restrict ourselves to considering equilibrium states, so the time derivative does not contribute either. Under these conditions, Eq. (2.3.30) becomes simply

$$\mathbf{T} = \mathbf{T}_1 + \mathbf{T}_2 = \mathbf{0}. \qquad (2.3.35)$$

In the argument that follows, we recall the vector identity

$$\nabla\,(\mathbf{a} \cdot \mathbf{b}) = (\mathbf{a}.\nabla)\,\mathbf{b} + (\mathbf{b} \cdot \nabla)\,\mathbf{a} + \mathbf{a} \times (\nabla \times \mathbf{b}) + \mathbf{b} \times (\nabla \times \mathbf{a}), \qquad (2.3.36)$$

and note that the constitutive relations are given by

$$\mathbf{D}(\mathbf{r}, \omega) = \epsilon\,(\mathbf{r}, \omega)\,\mathbf{E}(\mathbf{r}, \omega), \quad \mathbf{H}(\mathbf{r}, \omega) = \frac{1}{\mu\,(\mathbf{r}, \omega)}\mathbf{B}(\mathbf{r}, \omega), \qquad (2.3.37)$$

where we have chosen to absorb the constants $\epsilon_0$ and $\mu_0$ into the definitions of the respective fields. Under these conditions, and for a fixed frequency $\omega$,

$$
\begin{aligned}
(\nabla \times \mathbf{D}(\mathbf{r})) \times \mathbf{E}(\mathbf{r}) &= (\nabla \times \epsilon(\mathbf{r})\mathbf{E}(\mathbf{r})) \times \mathbf{E}(\mathbf{r}) \\
&= \nabla\epsilon(\mathbf{r})\,(\mathbf{E}(\mathbf{r}) \times \mathbf{E}(\mathbf{r})) + \epsilon(\mathbf{r})\,(\nabla \times \mathbf{E}(\mathbf{r})) \times \mathbf{E}(\mathbf{r}).
\end{aligned} \tag{2.3.38}
$$

Clearly, the first term, involving the cross product of **E** with itself, is identically zero. Evaluated at a fixed point $\mathbf{r}_0$,

$$
\begin{aligned}
[(\nabla \times \mathbf{D}(\mathbf{r})) \times \mathbf{E}(\mathbf{r})]_{\mathbf{r}_0} &= \epsilon(\mathbf{r}_0)\,[\nabla \times \mathbf{E}(\mathbf{r})]_{\mathbf{r}_0} \times \mathbf{E}(\mathbf{r}_0) \\
&= [\nabla \times \mathbf{E}(\mathbf{r})]_{\mathbf{r}_0} \times \epsilon(\mathbf{r}_0)\mathbf{E}(\mathbf{r}_0).
\end{aligned} \tag{2.3.39}
$$

Hence we may write

$$
[(\nabla \times \mathbf{D}(\mathbf{r})) \times \mathbf{E}(\mathbf{r})]_{\mathbf{r}_0} = [(\nabla \times \mathbf{E}\,(\mathbf{r})) \times \mathbf{D}(\mathbf{r})]_{\mathbf{r}_0}\,, \tag{2.3.40}
$$

in which the field terms have been exchanged, noting that

$$
\mathbf{D}(\mathbf{r}) = \epsilon(\mathbf{r})\mathbf{E}(\mathbf{r}). \tag{2.3.41}
$$

We may perform the same trick for the magnetic fields:

$$
[(\nabla \times \mathbf{H}(\mathbf{r})) \times \mathbf{B}(\mathbf{r})]_{\mathbf{r}_0} = [(\nabla \times \mathbf{B}\,(\mathbf{r})) \times \mathbf{H}(\mathbf{r})]_{\mathbf{r}_0}\,, \tag{2.3.42}
$$

noting that

$$
\mathbf{H}(\mathbf{r}) = \frac{1}{\mu(\mathbf{r})}\mathbf{B}(\mathbf{r}). \tag{2.3.43}
$$

In addition, we find that

$$
\begin{aligned}
(\mathbf{E}(\mathbf{r}) \cdot \nabla)\,\mathbf{D}(\mathbf{r}) &= \mathbf{E}(\mathbf{r}) \cdot \nabla\,(\epsilon(\mathbf{r})\mathbf{E}(\mathbf{r})) \\
&= \epsilon(\mathbf{r})\,(\mathbf{E}(\mathbf{r}) \cdot \nabla)\,\mathbf{E}(\mathbf{r}) + \nabla\epsilon(\mathbf{r})\,(\mathbf{E}(\mathbf{r}) \cdot \mathbf{E}(\mathbf{r}))\,.
\end{aligned} \tag{2.3.44}
$$

Evaluated at a point $\mathbf{r}_0$ :

$$
[(\mathbf{E}(\mathbf{r}) \cdot \nabla)\,\mathbf{D}]_{\mathbf{r}_0} = \Big[(\mathbf{D}(\mathbf{r}) \cdot \nabla)\,\mathbf{E}(\mathbf{r}) + \nabla\epsilon(\mathbf{r})E^2(\mathbf{r})\Big]_{\mathbf{r}=\mathbf{r}_0}\,. \tag{2.3.45}
$$

We note this time that, in exchanging the order of the fields, we acquire an additional term that depends upon the gradient of the permittivity. Similarly, we find

$$
[(\mathbf{B}(\mathbf{r}) \cdot \nabla)\,\mathbf{H}]_{\mathbf{r}_0} = \left[(\mathbf{H}(\mathbf{r}) \cdot \nabla)\,\mathbf{B}(\mathbf{r}) + \nabla\left(\frac{1}{\mu(\mathbf{r})}\right)B^2(\mathbf{r})\right]_{\mathbf{r}=\mathbf{r}_0}\,, \tag{2.3.46}
$$

where we have again acquired an additional term that depends upon the gradient of the permeability. Using Eq. (2.3.40) and (2.3.45) and the vector identity (2.3.36), the first term of $\mathbf{T}_1$, evaluated at $\mathbf{r}_0$, can be rewritten

$$(\nabla \times \mathbf{E}) \times \mathbf{D} = (\mathbf{D} \cdot \nabla)\,\mathbf{E} - \frac{1}{2}\nabla\,(\mathbf{D} \cdot \mathbf{E}) + \frac{1}{2}\nabla\,[\epsilon(\mathbf{r})]\,E^2(\mathbf{r}). \tag{2.3.47}$$

Similarly, for the third term, using Eqs. (2.3.35), (2.3.42) and (2.3.46), we obtain

$$(\nabla \times \mathbf{H}) \times \mathbf{B} = (\mathbf{B}.\nabla)\,\mathbf{H} - \frac{1}{2}\nabla\,(\mathbf{B} \cdot \mathbf{H}) + \frac{1}{2}\nabla\left[\frac{1}{\mu(\mathbf{r})}\right] B^2(\mathbf{r}). \tag{2.3.48}$$

Reexpressing (2.3.33) using (2.3.42–2.3.48), it follows from (2.3.35) that

$$\nabla \cdot \sigma_M = -\frac{1}{2}\left(\nabla\,[\epsilon(\mathbf{r})]\,E^2(\mathbf{r}) + \nabla\left[\mu(\mathbf{r})^{-1}\right] B^2(\mathbf{r})\right), \tag{2.3.49}$$

where $\sigma_M$ defines the familiar Minkowski stress tensor for the electromagnetic field

$$\sigma_M = (\mathbf{D} \otimes \mathbf{E}) + (\mathbf{B} \otimes \mathbf{H}) - \frac{1}{2}\,(\mathbf{D} \cdot \mathbf{E} + \mathbf{B} \cdot \mathbf{H})\,\mathbf{1}_3. \tag{2.3.50}$$

In fact, expression (2.3.49) is equal to $\mathbf{f}_{\mathcal{E}}$, defined in Eq. (2.3.18). It follows from the right-hand side of (2.3.49) that, in an equilibrium state, in the absence of external currents or charges, $\sigma_M$ may be alternatively cast in the form

$$\sigma_{\mathrm{M}} = -\frac{1}{2}\left\{\int \nabla\epsilon(\mathbf{r})E^2(\mathbf{r}) \cdot \mathrm{d}\mathbf{r} + \int \nabla\left(\frac{1}{\mu\,(\mathbf{r})}\right) B^2(\mathbf{r}) \cdot \mathrm{d}\mathbf{r}\right\} \mathbf{1}_3. \tag{2.3.51}$$

#### 2.3.1.4 The Abraham Stress Tensor

Exchanging $\sigma_{\mathcal{E}}$ for $\sigma_M$ in the stress tensor (2.3.21), and rewriting $\sigma_M$ in the form

$$\sigma_M = (\mathbf{D} \otimes \mathbf{E}) + (\mathbf{B} \otimes \mathbf{H}) - \frac{1}{2}\left(\epsilon E^2 + \frac{1}{\mu}B^2\right)\mathbf{1}_3, \tag{2.3.52}$$

we recover the celebrated Abraham tensor of the stationary electromagnetic field in a dielectric fluid:

$$\sigma = -P_0\mathbf{1}_3 + (\mathbf{D} \otimes \mathbf{E}) + (\mathbf{B} \otimes \mathbf{H}) - \frac{1}{2}\left\{E^2\left(\epsilon - \rho\frac{\partial\epsilon}{\partial\rho}\right) + B^2\left(\frac{1}{\mu} - \rho\frac{\partial}{\partial\rho}\frac{1}{\mu}\right)\right\}\mathbf{1}_3. \tag{2.3.53}$$

This is the total stress tensor applied in the general form of Lifshitz theory [12] for the prediction of Casimir phenomena in media.

#### 2.3.1.5 The Casimir-Lifshitz Force and the Fluid Pressure

The Casimir-Lifshitz force in a liquid medium may be understood as arising in the equilibrium balance between the radiation pressure and fluid pressure components of the stress tensor. Its stress tensor is given by the quantum analogue of

$$\sigma_M = (\mathbf{D} \otimes \mathbf{E}) + (\mathbf{B} \otimes \mathbf{H}) - \frac{1}{2} (\mathbf{D} \cdot \mathbf{E} + \mathbf{B} \cdot \mathbf{H})\, \mathbf{1}_3, \tag{2.3.54}$$

which may be associated with an electromagnetic force density

$$\mathbf{f}_C = \mathbf{f}_M = \nabla \cdot \sigma_M. \tag{2.3.55}$$

It may alternatively be expressed in the form

$$\mathbf{f}_C = \mathbf{f}_{\mathcal{E}} = \nabla \cdot \sigma_{\mathcal{E}} \equiv -\frac{1}{2} \left( \nabla \left[\epsilon(\mathbf{r})\right] E^2(\mathbf{r}) + \nabla \left[\mu(\mathbf{r})^{-1}\right] B^2(\mathbf{r}) \right), \tag{2.3.56}$$

where it is specified in terms of the gradients of the dielectric functions. It is exactly opposed by a compensating stress containing the pressure gradient due to the fields,

$$\sigma_P = \left\{ -P_0 + \frac{1}{2} \left( \rho \frac{\partial \epsilon}{\partial \rho} E^2 + \rho \frac{\partial}{\partial \rho} \left( \frac{1}{\mu} \right) B^2 \right) \right\} \mathbf{1}_3. \tag{2.3.57}$$

Consequently,

$$\nabla \cdot \sigma = \nabla \cdot \sigma_M + \nabla \cdot \sigma_P = \mathbf{f}_C + \mathbf{f}_P = 0, \tag{2.3.58}$$

and mechanical equilibrium is maintained. The Casimir force density may thus be inferred either from the quantum analogues of the pressure or Minkoswki contributions to the stress:

$$\mathbf{f}_C = -\mathbf{f}_P. \tag{2.3.59}$$

The total pressure in the fluid (a scalar quantity) includes the contribution $P_0$ defined in the absence of electromagnetic fluctuations.

#### 2.3.1.6 The Casimir Force on Interacting Bodies in a Medium

The Casimir-Lifshitz force is the quantum analogue of the force determined by the volume integral of the force density, which is equivalent to the integration of the Minkowski stress over an enclosing surface:

$$\mathbf{F}_C = \int_V \mathrm{d}\mathbf{r}\, \nabla \cdot \sigma_M = \int_{\partial V} \mathrm{d}\mathbf{A} \cdot \sigma_M. \tag{2.3.60}$$

From Eq. (2.3.56), we observe that $\nabla \cdot \sigma_M = 0$ and hence $\mathbf{F}_C = 0$ for any surface enclosing a volume of uniform fluid. However, for any surface surrounding a solid body, the gradients of the dielectric functions are non-zero, and integration over the opposing electromagnetic or pressure components of the stress produces the force acting on the body.

#### 2.3.1.7 Piece-wise Homogeneous Case

Let us consider now a piece-wise planar homogeneous case with a boundary at $x = x_0$, as a special case of an inhomogeneous system. In fact, most theoretical Casimir problems are concerned with the interactions between homogeneous bodies; consequently, their global dielectric functions are piece-wise homogeneous. The constitutive relations are as stated earlier, but here $\epsilon$ is sharply discontinuous at the boundary:

$$\epsilon\,(x) = \begin{cases} \epsilon_1 & x < x_0 \\ \epsilon_2 & x \geq x_0 \end{cases}. \tag{2.3.61}$$

The permittivity $\epsilon(x)$ can be rewritten in terms of the Heaviside function $H$:

$$\epsilon(x) = \epsilon_1 + H(x - x_0)\Delta\epsilon, \tag{2.3.62}$$

where $\Delta\epsilon = \epsilon_2 - \epsilon_1$. The magnetic response of a material can often be ignored ($\mu \to 1$). Consider Eq. (2.3.56). For the piece-wise homogeneous system of our example, the first term involves

$$\frac{\partial}{\partial x}\left[\epsilon(x)\right] E^2(\mathbf{r}) = \delta(x - x_0)\,\Delta\epsilon\, E^2(\mathbf{r}). \tag{2.3.63}$$

The second term also gives a non-zero contribution only at $x = x_0$, i.e. on the boundary. For piece-wise homogeneous media, then, the stress is discontinuous at the interface between two different media, and the force-density is concentrated in a delta-function (in this case, at $x = x_0$):

$$\nabla \cdot \sigma_M = -\frac{1}{2}\delta(x - x_0)\left[\Delta\epsilon\, E^2(\mathbf{r})\right]. \tag{2.3.64}$$

Notice that inside an infinite homogeneous medium, where the gradients of the dielectric functions are zero, Eq. (2.3.56) implies that the Casimir force must be identically zero.

#### 2.3.1.8 General Remarks

We find then that the force density can be written solely in terms containing spatial derivatives of the dielectric functions, for both piecewise homogeneous systems and inhomogeneous media. Rewritten as a quantum correlation function, averaged over the vacuum state, $\sigma_M$ ultimately affords the Casimir force density for a system, including Casimir's original case involving a cavity formed by two perfect mirrors. The Casimir force therefore depends upon material inhomogeneities in a system that scatter the (virtual) photons of the electromagnetic field, and it is this scattered part of the field that is relevant to computing the Casimir stresses in a system. This is a thought we will repeatedly come back to.

### *2.3.2 Averaging over the Quantum Stress Tensor*

We have established that the Casimir force density is equal to the divergence of the Minkowski stress tensor. Of course, this tensor must first be quantised. The work has essentially been done, however: it is sufficient to replace the classical field components of the stress tensor with the quantum operators we have defined in this chapter:

$$\hat{\sigma} = \hat{\mathbf{E}} \otimes \hat{\mathbf{D}} + \hat{\mathbf{B}} \otimes \hat{\mathbf{H}} - \frac{1}{2}\left(\hat{\mathbf{E}} \cdot \hat{\mathbf{D}} + \hat{\mathbf{B}} \cdot \hat{\mathbf{H}}\right) \mathbf{1}_3. \tag{2.3.65}$$

This effectively rewrites the stress in terms of the fundamental field operators. Using the Gauss theorem, the total force operator can then be written in the form

$$\hat{\mathbf{F}} = \int\limits_{\partial V} \mathrm{d}\mathbf{A} \cdot \hat{\sigma} - \int\limits_{V} \mathrm{d}^3 r \frac{\partial}{\partial t}\left(\hat{\mathbf{D}} \times \hat{\mathbf{B}}\right). \tag{2.3.66}$$

To obtain the measured Casimir force, we must average the stress tensor in the ground state of the polariton field. The force is then

$$\mathbf{F} = \int\limits_{\partial V} \mathrm{d}\mathbf{A} \cdot \langle\{0\}|\, \hat{\sigma}\, |\{0\}\rangle - \int\limits_{V} \mathrm{d}^3 r \frac{\partial}{\partial t} \langle\{0\}|\, \hat{\mathbf{D}} \times \hat{\mathbf{B}}\, |\{0\}\rangle. \tag{2.3.67}$$

According to the Schrödinger equation, the ground-state is constant in time. The second term therefore vanishes, and we can state the force as

$$\mathbf{F} = \int\limits_{\partial V} \mathrm{d}\mathbf{A} \cdot \sigma, \tag{2.3.68}$$

where $\sigma$ is the expectation value of the stress tensor in the ground-state. The ground-state expectation values can be computed by expanding the fields in terms of the fundamental bosonic field operators (2.2.31–2.2.36), exploiting their properties (2.2.23), but neglecting absorption.[9] In doing so, we find that [1, 16]

$$\left\langle \hat{\mathbf{E}}\left(\mathbf{r}\right) \otimes \hat{\mathbf{D}}\left(\mathbf{r}'\right)\right\rangle = \frac{\hbar}{\pi} \int_0^\infty \mathrm{d}\omega \mu_0 \omega^2 \epsilon_0 \epsilon\left(\mathbf{r}, \omega\right) \operatorname{Im} \mathbf{G}(\mathbf{r}, \mathbf{r}', \omega), \tag{2.3.69}$$

$$\left\langle \hat{\mathbf{B}}\left(\mathbf{r}\right) \otimes \hat{\mathbf{H}}\left(\mathbf{r}'\right)\right\rangle = \frac{\hbar}{\pi} \int_0^\infty \mathrm{d}\omega \frac{1}{\mu(\mathbf{r}, \omega)} \nabla \times \operatorname{Im} \mathbf{G}(\mathbf{r}, \mathbf{r}', \omega) \times \overleftarrow{\nabla'} . \tag{2.3.70}$$

The stress tensor, evaluated at position $\mathbf{r}$, is then readily cast into the form [1]

$$\sigma\left(\mathbf{r}\right) = \frac{\hbar}{\pi} \int_0^\infty \mathrm{d}\omega \left(\tau\left(\mathbf{r}\right) - \frac{1}{2}\operatorname{Tr}\left[\tau\left(\mathbf{r}\right)\right] \mathbf{1}_3\right), \tag{2.3.71}$$

where Tr implements the trace function. This involves integrating over all frequencies the function

$$\tau\left(\mathbf{r}\right) = \frac{\omega^2}{c^2} \epsilon\left(\mathbf{r}, \omega\right) \operatorname{Im} \mathbf{G}(\mathbf{r}, \mathbf{r}, \omega) + \frac{1}{\mu(\mathbf{r}, \omega)} \left[\nabla \times \operatorname{Im} \mathbf{G}(\mathbf{r}, \mathbf{r}', \omega) \times \overleftarrow{\nabla'}\right]_{\mathbf{r}'=\mathbf{r}} . \tag{2.3.72}$$

The stress is therefore determined by the classical Green function of the field.

### 2.3.3 Regularising the Stress Tensor

However, the stress tensor (2.3.71), like the zero-point energy, turns out to be *infinite*. To facilitate the removal of the divergence, we introduce an additional positional argument $\mathbf{r}'$ and define the regularised stress tensor

$$\sigma = \lim_{\mathbf{r}' \to \mathbf{r}} \left\langle \hat{\sigma}\left(\mathbf{r}, \mathbf{r}'\right) - \hat{\sigma}_0(\mathbf{r}, \mathbf{r}')\right\rangle . \tag{2.3.73}$$

The first term of (2.3.73) is a quantum correlation function defined by

$$\hat{\sigma}\left(\mathbf{r}, \mathbf{r}'\right) = \hat{\mathbf{E}}(\mathbf{r}) \otimes \hat{\mathbf{D}}(\mathbf{r}') + \hat{\mathbf{B}}(\mathbf{r}) \otimes \hat{\mathbf{H}}(\mathbf{r}') - \frac{1}{2}\left(\hat{\mathbf{E}}(\mathbf{r}) \cdot \hat{\mathbf{D}}(\mathbf{r}') + \hat{\mathbf{B}}(\mathbf{r}) \cdot \hat{\mathbf{H}}(\mathbf{r}')\right) \mathbf{1}_3, \tag{2.3.74}$$

---

[9] We retain only the *real* parts. To take proper account of absorption in media requires a more sophisicated formulation of macro-QED [4]. However, the stress tensor is the same [3].

which is identical in form to the quantum stress in the limit as $\mathbf{r}' \to \mathbf{r}$. The subtraction of the second term $\hat{\sigma}_0$ in (2.3.73) is typically associated with the removal of illicit 'self-forces'. Both of these terms in the expression are finite for $\mathbf{r}' \neq \mathbf{r}$, so the subtraction is well-defined. Applying once again results (2.3.69) and (2.3.70), the stress correlation function (2.3.74) can be expressed in the form [1]

$$\sigma(\mathbf{r}, \mathbf{r}') = -\frac{\hbar}{\pi} \int_0^\infty \mathrm{d}\xi \left( \tau(\mathbf{r}, \mathbf{r}') - \frac{1}{2} \mathrm{Tr} \left[ \tau(\mathbf{r}, \mathbf{r}') \right] \mathbf{1}_3 \right), \tag{2.3.75}$$

which involves integrating over the imaginary axis the associated correlation function[10]

$$\tau(\mathbf{r}, \mathbf{r}') = \frac{\xi^2}{c^2} \epsilon(\mathbf{r}, i\xi)\, \mathbf{G}(\mathbf{r}, \mathbf{r}', i\xi) - \frac{1}{\mu(\mathbf{r}, i\xi)} \nabla \times \mathbf{G}(\mathbf{r}, \mathbf{r}', i\xi) \times \overleftarrow{\nabla}' . \tag{2.3.76}$$

The additional step of rotating to the imaginary axis is a mathematical trick to produce a more rapidly converging integral, necessitating the evaluation of the dielectric functions at imaginary frequencies. These quantities can be obtained from the original functions by Hilbert transforms. The correlation function is expressed in terms of the classical Green function of the electromagnetic field, which may be associated in this case with the magnitude of a field at $\mathbf{r}$ produced by a dipole situated at $\mathbf{r}'$ oscillating with 'imaginary frequency' $\xi$, which probes the properties of the system.

#### 2.3.3.1 The Meaning of the Regularisation

Concerning the regularisation procedure, it must be emphasised that, in attempting to isolate and remove the divergences in the stress tensor, we are not in general at liberty to subtract anything that depends upon the inhomogeneity of the medium. This could abitrarily modify the Casimir force, which itself arises as a result of scattering due to material inhomogeneities in the system (see Eq. (2.3.56)), as we have discussed. This fairly obvious stricture has frequently been violated without adequate justification in the literature. However, the purchase of finitude at the cost of *meaning* is too high a price to pay. In Lifshitz theory, the regularisation is assigned a fairly precise semantics [14] best expressed at the level of the Green function, which we may write as

$$\mathbf{G}(\mathbf{r}, \mathbf{r}') = \left( \mathbf{G}(\mathbf{r}, \mathbf{r}') - \mathbf{G}_0(\mathbf{r}, \mathbf{r}') \right) + \mathbf{G}_0(\mathbf{r}, \mathbf{r}'), \tag{2.3.77}$$

where $\mathbf{G}_0$ is an auxiliary Green function for an infinite homogeneous medium whose dieletric properties are the same as that of the actual medium at the point $r'$. As we have argued, there is no Casimir force in an infinite homogeneous medium; the third term in (2.3.77) can effectively be absorbed into a bulk stress term that is not associ-

[10] For a discussion of Wick rotation to the imaginary axis, see Appendix A.

ated with the Casimir force. This results in a renormalisation of the stress in which both the form and meaning of $\sigma_0$ in Eq. (2.3.73) are now apparent: to determine the *physical* part of the stress tensor relevant to the measurable Casimir force, we subtract a bulk stress $\sigma_0$ corresponding to an infinite homogeneous medium, effectively renormalising the Casimir stress to zero in the absence of any inhomogeneity in the system. We may also define the renormalised Green function

$$\mathbf{G}_S(\mathbf{r}, \mathbf{r}') = \mathbf{G}(\mathbf{r}, \mathbf{r}') - \mathbf{G}_0(\mathbf{r}, \mathbf{r}'), \tag{2.3.78}$$

corresponding to the *scattered* part of the electromagnetic field. Simply computing the stress using the scattered Green function (2.3.78) is equivalent to computing the stress using Eq. (2.3.73).

## 2.4 The Lifshitz Result for Two Half-spaces

Lifshitz' original paper considered the case of the force produced due to fluctuations between two parallel dielectric half-spaces, separated by empty space [5]. This result was subsequently generalised to allow for the case of a liquid medium between the moving bodies, as well as vacuum, producing a general expression for the force that recovers Casimir's own expression in appropriate limits [12]. By solving the wave equation to determine the correct Green function, we can recover this well-known result from our more general expressions (2.3.73–2.3.76). The calculation is somewhat involved, however, and we present here only an outline.

### *2.4.1 The Force in a Dielectric Sandwich*

#### 2.4.1.1 The Planar Green Function

In planar dielectrics, we can arrange our coordinate system so that $\varepsilon$ and $\mu$ depend on $x$, but not on $y$ and $z$. In view of the symmetry in $y$ and $z$, we introduce a spatial Fourier transformation to simplify the problem:

$$G(\mathbf{r}, \mathbf{r}') = \frac{1}{(2\pi)^2} \int_{-\infty}^{+\infty}\int_{-\infty}^{+\infty} \tilde{G}(x, x', k_y, k_z)\, e^{ik_y(y-y')+ik_z(z-z')} \mathrm{d}k_y \mathrm{d}k_z, \tag{2.4.1}$$

noting that

$$\nabla = (\partial_x, ik_y, ik_z), \quad \nabla' = (\partial_{x'}, -ik_y, -ik_z). \tag{2.4.2}$$

The Green function obeys the Fourier transformed wave equation

$$\nabla \times \frac{1}{\mu} \nabla \times \tilde{G} + \varepsilon \kappa^2 \tilde{G} = \mathbf{1}_3 \delta(x - x'), \tag{2.4.3}$$

where $\kappa = \xi / c$. Solving the wave equation is slightly involved, even for this simple system, and we will not repeat all the details here.[11] The Fourier-transformed Green function may be expressed in the form

$$\tilde{G} = \tilde{G}_e + \tilde{G}_m, \quad \text{for } x \neq x', \tag{2.4.4}$$

consisting of two separate contributions [1],

$$\tilde{G}_e = \mathbf{n}_E \tilde{g}_e \otimes \mathbf{n}'_E, \quad \tilde{G}_m = -\frac{\nabla \times \mathbf{n}_E \tilde{g}_m \otimes \mathbf{n}'_E \times \overleftarrow{\nabla'}}{\varepsilon \varepsilon' \kappa^2}, \tag{2.4.5}$$

where

$$\mathbf{n}_E = \frac{1}{\sqrt{k_y^2 + k_z^2}} \begin{pmatrix} 0 \\ -k_z \\ k_y \end{pmatrix}, \quad \mathbf{n}'_E = -\mathbf{n}_E. \tag{2.4.6}$$

Primed terms are evaluated at position $r'$, and the operator $\overleftarrow{\nabla'}$ acts to the left with respect to $r'$. For this geometry, the problem may be essentially rewritten in terms of electric and magnetic scalar Green functions that solve the wave equations

$$\nabla \cdot \frac{1}{\mu(x)} \nabla \tilde{g}_e(x) - \varepsilon(x) \kappa^2 \tilde{g}_e(x) = \delta(x - x'), \tag{2.4.7}$$

$$\nabla \cdot \frac{1}{\epsilon(x)} \nabla \tilde{g}_m(x) - \mu(x) \kappa^2 \tilde{g}_m(x) = \delta(x - x'). \tag{2.4.8}$$

#### 2.4.1.2 The Planar Stress

We can write the stress correlation functions (2.3.75, 2.3.76) in terms of the Fourier-transformed Green function (2.4.4), shifting the prefactor and the integations to the second correlation function:

$$\sigma = \sum_{\lambda = e,m} \left( \tau_\lambda - \frac{1}{2} \mathrm{Tr}\, \tau_\lambda\, \mathbf{1}_3 \right), \tag{2.4.9}$$

where we have separated the stress into components that depend on the electric and magnetic Green functions, and

---

[11] The full details of this approach can be found in [1].

$$\tau_\lambda = \frac{\hbar c}{\pi} \int_0^\infty d\kappa \int_{-\infty}^{+\infty} \frac{d\mathbf{k}_\parallel}{(2\pi)^2} \left( -\kappa^2 \varepsilon \tilde{G}_\lambda + \frac{1}{\mu} \nabla \times \tilde{G}_\lambda \times \overleftarrow{\nabla'} \right), \tag{2.4.10}$$

with $\mathbf{k}_\parallel = (0, k_y, k_z)$. The inverse Fourier-transform nullifies any terms that are odd in $k_y$ or $k_z$ and both the $\tilde{G}_e$ and the $\tilde{G}_m$ contributions have 'off-diagonal' components that then vanish in the integral. It follows that the stress correlation function is isotropic, i.e.

$$\sigma = \mathrm{diag}(\sigma_{xx}, \sigma_{yy}, \sigma_{zz}). \tag{2.4.11}$$

Furthermore, the Green function does not depend on $y$ and $z$, which is evident from the symmetry of this problem. The only component that is relevant to the Casimir force, in this case, is $\sigma_{xx}$, where

$$\sigma_{xx} = \sum_{\lambda=e,m} \left( \tau_{\lambda,xx} - \frac{1}{2} \mathrm{Tr}\, \tau_\lambda \right). \tag{2.4.12}$$

Consider the contribution of the electric component of the correlation function. We find that

$$\tau_{e,xx} = -\frac{\hbar c}{\pi} \int_0^\infty d\kappa \int_{-\infty}^{+\infty} \frac{d\mathbf{k}_\parallel}{(2\pi)^2} \frac{1}{\mu} (k_y^2 + k_z^2) \tilde{g}_e, \tag{2.4.13}$$

$$\mathrm{Tr}\, \tau_e = \frac{\hbar c}{\pi} \int_0^\infty d\kappa \int_{-\infty}^{+\infty} \frac{d\mathbf{k}_\parallel}{(2\pi)^2} \left( \kappa^2 \varepsilon - \frac{1}{\mu} (k_y^2 + k_z^2 + \partial_x \partial_{x'}) \right) \tilde{g}_e. \tag{2.4.14}$$

It follows that the electric component of the stress is of the form [1]

$$\sigma_{e,xx} = -\frac{\hbar c}{2} \int_0^\infty d\kappa \int_{-\infty}^{+\infty} \frac{d\mathbf{k}_\parallel}{(2\pi)^2} \frac{1}{\mu} (w^2 - \partial_x \partial_{x'}) \tilde{g}_e, \tag{2.4.15}$$

where

$$w = \sqrt{k_y^2 + k_z^2 + n^2 \kappa^2}, \quad n = \sqrt{\varepsilon \mu}. \tag{2.4.16}$$

Similarly, for the magnetic contribution, we find that

$$\sigma_{m,xx} = -\frac{\hbar c}{2} \int_0^\infty d\kappa \int_{-\infty}^{+\infty} \frac{d\mathbf{k}_\parallel}{(2\pi)^2} \frac{1}{\varepsilon} (w^2 - \partial_x \partial_{x'}) \tilde{g}_m. \tag{2.4.17}$$

Both of these contributing terms, however, are infinite. To obtain a finite and physical result, we need to renormalise the stress tensor. This may be achieved at the level of the scalar Green functions by isolating the scattered components of $\tilde{g}_e$ and $\tilde{g}_m$.

#### 2.4.1.3 The Scattered Green Function

We are considering a dielectric sandwich consisting of three uniform regions characterised by dielectric constants $\{\varepsilon_1, \mu_1\}$, $\{\epsilon_2, \mu_2\}$, and $\{\varepsilon_3, \mu_3\}$. The outer dielectrics are idealised to be infinitely thick, the left outer plate extending from $x \in (-\infty, 0)$, the inner region from $x \in [0, a]$, and the right outer plate from $x \in (a, \infty)$. We could proceed by solving the scalar wave equations, with the corresponding boundary conditions, to obtain $\tilde{g}_e$ and $\tilde{g}_m$. However, it is mathematically simpler (and physically more engaging) to determine the solution by thinking about the structure of the scalar waves and the multiple propagations and reflections that will occur at the boundaries.

The scalar Green functions should describe waves emitted at $x'$ that are multiply reflected at the dielectric interfaces. Let $r_L$ and $r_R$ be the left and right reflection coefficients of the plates. These factors are determined by the relevant permittivities and permeabilities. Multiple reflections and propagations add up to

$$\sum_{m=0}^{\infty} \left( \mathrm{e}^{-2aw} r_L r_R \right)^m = \frac{1}{1 - \mathrm{e}^{-2aw} r_L r_R}. \tag{2.4.18}$$

Now consider all possible multiple reflections and propagations from $x'$ with the strength $-1/2w$ of the bare Green function

$$\tilde{g}_0 = -\frac{1}{2w} \mathrm{e}^{(-w|x-x'|)}. \tag{2.4.19}$$

This collection consists of forward and backwards-directed waves emitted at $x'$ and measured at $x$, after multiple reflections. For the forwards-directed waves, there are two cases:

1. when $x' < x$, the wave proceeds directly to $x$ on its first cycle, and the distance between the source and the point of measurement is simply $x - x'$;
2. when $x' > x$, the wave picks up an extra reflection $r_R$ on the right on its first cycle; the distance between source and measurement is $(a - x') + (a - x) = 2a - x - x'$.

And for the backwards-directed waves, there are also two cases:

1. when $x' > x$, the wave proceeds directly to $x$ on its first cycle; the distance between source and measurement is $x' - x$.
2. when $x' < x$, the wave picks up an extra reflection $r_L$ on the left before it is measured at $x$; the distance between source and measurement is $x' + x' + (x - x') = x + x'$.

In total, this gives

$$\tilde{g}_1 = -\frac{1}{\left(1 - \mathrm{e}^{-2aw} r_L r_R\right)} \cdot \frac{1}{2w} \left(\mathrm{e}^{-w(x-x')} + \mathrm{e}^{-w(x+x')} r_L + \mathrm{e}^{w(x-x')} + \mathrm{e}^{w(x+x'-2a)} r_R\right). \tag{2.4.20}$$

Notice, however, that if $x' < x$, the direct propagation from $x'$ to $x$ for the backwards-directed wave,

$$\tilde{g} = -\frac{1}{2w} \mathrm{e}^{w(x-x')}, \tag{2.4.21}$$

is not possible; only indirect propagations that involve multiple reflections are possible. Likewise, if $x < x'$, then direct forward-propagating waves are not possible. We must therefore subtract these contributions[12]; *ergo*, we must add the scalar Green function

$$\tilde{g}_2 = \frac{1}{2w} \left(\mathrm{e}^{w(x-x')} + \mathrm{e}^{-w(x-x')}\right). \tag{2.4.22}$$

We may therefore express the scalar Green function as

$$\tilde{g} = \tilde{g}_0 + \tilde{g}_1 + \tilde{g}_2. \tag{2.4.23}$$

The renormalised scalar Green function, which is associated with scattering, is then

$$\tilde{g}_S = \tilde{g} - \tilde{g}_0 = \tilde{g}_1 + \tilde{g}_2. \tag{2.4.24}$$

Consider that, in the stress function, the terms $\mathrm{e}^{-w(x+x')} r_L$ and $\mathrm{e}^{w(x+x'-2a)} r_R$ are eliminated due to the term $(w^2 - \partial_x \partial_{x'})\tilde{g}_S$. The only terms that are retained are those that contain $\mathrm{e}^{\pm w(x-x')}$, and these are effectively doubled. Incorporating this factor of two into the stress, and taking the limit $x' \to x$, we can express the scalar Green function as [1]

$$\tilde{g}_S = -\frac{1}{w(r_L^{-1} r_R^{-1} \mathrm{e}^{2aw} - 1)}. \tag{2.4.25}$$

#### 2.4.1.4 The Regularised Stress

It is now possible to state the relevant, renormalised component of the stress exactly:

$$\sigma_{xx} = 2\hbar c \sum_{\lambda=e,m} \int_0^{\infty} \frac{\mathrm{d}\kappa}{2\pi} \int_{-\infty}^{+\infty} \frac{\mathrm{d}^2 \mathbf{k}_\parallel}{(2\pi^2)} w \frac{r_{\lambda L} r_{\lambda R}\, \mathrm{e}^{-2aw}}{1 - r_{\lambda L} r_{\lambda R}\, \mathrm{e}^{-2aw}}. \tag{2.4.26}$$

---

[12] In the limit as $x' \to x$, this amounts to subtracting the self-interacting terms. Only waves which have explored the environment contribute to the Casimir force.

This is Lifshitz' celebrated result [5] in a more general form.[13] When we consider the Green functions in the outer plates, we note that the only terms that depend on $e^{\pm w(x-x')}$ are those describing direct propagation, and that these are subtracted from $\tilde{g}$ in $\tilde{g}_S$. It follows that the Casimir force vanishes in the outer plates. The force density is concentrated in the interface as a delta function, pointing inwards (attractive) if $\sigma_{xx}$ is positive and outward (repulsive) if $\sigma_{xx}$ is negative. For the case of uniform plates, the electromagnetic reflection coefficients are given by [2]:

$$r_{eL} = \frac{\mu_1 w_2 - \mu_2 w_1}{\mu_1 w_2 + \mu_2 w_1}, \quad r_{mL} = \frac{\varepsilon_1 w_2 - \varepsilon_2 w_1}{\varepsilon_1 w_2 + \varepsilon_2 w_1}, \tag{2.4.27}$$

$$r_{eR} = \frac{\mu_3 w_2 - \mu_2 w_3}{\mu_3 w_2 + \mu_2 w_3}, \quad r_{mR} = \frac{\varepsilon_3 w_2 - \varepsilon_2 w_3}{\varepsilon_3 w_2 + \varepsilon_2 w_3}, \tag{2.4.28}$$

where the physical quantities are evaluated in the subscripted regions.

### 2.4.2 The Casimir Force in the Limit

To recover Casimir's original result, we assume vacuum between the plates $\varepsilon_2 = \mu_2 = 1$, and we take the limit $\varepsilon_1, \varepsilon_3 \to \infty$, with $\mu_l, \mu_r = 1$. The second condition corresponds to the case of a perfect electric mirror. We see from (2.4.27) that $r_{eL} = -1$. Similarly, $r_{eR} = r_{mL} = r_{mR} = -1$. This is the phase jump of $\pi$ at the mirror. We obtain

$$\sigma_{xx} = \frac{\hbar c}{2\pi^3} \int_0^{\infty} \int_{-\infty}^{+\infty} \int_{-\infty}^{+\infty} \frac{w}{e^{2aw} - 1} \mathrm{d}k_y \, \mathrm{d}k_z \, \mathrm{d}\kappa. \tag{2.4.29}$$

This integral is best expressed in spherical coordinates, with $w$ as the radius:

$$\sigma_{xx} = \frac{\hbar c}{2\pi^3} \int_0^{\infty} \int_0^{2\pi} \int_0^{\pi/2} \frac{w^3 \sin\phi}{e^{2aw} - 1} \mathrm{d}\phi \, \mathrm{d}\theta \, \mathrm{d}\kappa = \frac{\hbar c}{\pi^2} \int_0^{\infty} \frac{w^3}{e^{2aw} - 1}. \tag{2.4.30}$$

On calculating the integral, we find that

$$\sigma_{xx} = \frac{\hbar c \pi^2}{240 a^4} = \sigma_{\text{Casimir}}. \tag{2.4.31}$$

We have recovered Casimir's original result (1.1.23) from Lifshitz theory, expressed in terms of the Minkoswki stress.

[13] The Lifshitz result was not originally cast in terms of reflection coefficients, until Kats observed that it was natural to do so [26]. Indeed, expressed in this form [1] it remains valid even when the optical response of the mirrors cannot be described by a local dielectric response function [27].

### 2.4.3 Boyer's Repulsive Mirrors

As a passing curiosity that has prompted much speculation, there is a second and similar case that has been calculated [17] in which the Casimir force turns out to be repulsive. This time, whilst retaining a perfect electric mirror on the left, with $\varepsilon_1 \to \infty$, $\mu_1 = 1$, we substitute a perfect magnetic mirror on the right, with $\varepsilon_3 = 1$, $\mu_3 \to \infty$. In this case, the reflection coefficients are $r_{eL} = r_{mL} = -1$, but $r_{eR} = r_{mR} = 1$. It follows that

$$\sigma_{xx} = -\frac{\hbar c}{2\pi^3} \int_0^{\infty} \int_{-\infty}^{+\infty} \int_{-\infty}^{+\infty} \frac{w}{e^{2aw} + 1} dk_y\, dk_z\, d\kappa = -\frac{\hbar c}{\pi^2} \int_0^{\infty} \frac{w^3}{e^{2aw} + 1}. \tag{2.4.32}$$

On computing the integral, we find

$$\sigma_{xx} = -\frac{7}{8}\sigma_{\text{Casimir}}. \tag{2.4.33}$$

Here, the vacuum force causes an electric and a magnetic mirror to *repel* each other. However, like Casimir's result, this holds only for the highly idealised case of perfect mirrors. Repulsive Casimir forces in real materials remain the debated exception, with prohibitive theorems extending both to metamaterials[14] and chiral materials [18–20].

## References

1. U. Leonhardt, *Essential Quantum Optics* (Cambridge University Press, Cambridge, 2010)
2. J.D. Jackson, (Wiley, New York, 1998)
3. T.G. Philbin, New. J. Phys. **13**, 063026 (2011)
4. T.G. Philbin, New J. Phys. **12**, 123008 (2010)
5. E.M. Lifshitz, Zh. Eksp. Teor. Fiz. **29**, 94 (1955)
6. S.M. Rytov, V.I. Kravtsov, Y.A. Tatarskii (eds.), *Principles of Statistical Radiophysics*. Wave Propagation Through Random Media. vol. 3 (Springer, Berlin, 1989)
7. F.S.S Rosa, D.A.R Dalvit, P.W Milonni, Phys. Rev. A **81**, 033812 (2010)
8. S.Y. Buhmann, D.-G. Welsch, Prog. Quantum Electr. **31**(2), 51–130 (2007)
9. S.Y. Buhmann, Acta Phys. Slovaca **58**(5), 675–809 (2008)
10. S.Y. Buhmann, *Dispersion Forces I*. (Springer, Heidelberg, 2013)
11. R. Kubo, Rep. Prog. Phys. **29**(1), 255 (1966)
12. I.E. Dzyaloshinskii, E.M. Lifshitz, L.P. Pitaevskii, Adv. Phys. **10**, 165 (1961)
13. L.P. Pitaevskii, Phys. Rev. A **73**, 047801 (2006)
14. L.P. Pitaevskii, in *Casimir Physics*, ed. by D. Dalvit, P. Milonni, D. Roberts, F. da Rosa. Lecture Notes in Physics, vol 834 (Springer, Berlin, 2011), pp. 23-37
15. L.D. Landau, E.M. Lifshitz, *Electrodynamics of Continuous Media* (Pergamon Press, Oxford, 1960)
16. T.G. Philbin, C. Xiong, U. Leonhardt, Ann. Phys. **325**, 579 (2009)

[14] Metamaterials incorporate arrays of micro-engineered circuitry, and can be engineered to produce a strong magnetic response at certain frequencies.

17. T.H. Boyer, Phys. Rev. A **9**, 2078 (1974)
18. F.S.S. Rosa, J. Phys. Conf. Ser. **161**(1), 012039 (2009)
19. S.J. Rahi, T. Emig, Phys. Rev. Lett. **105**, 070404 (2010)
20. S.J. Rahi, T. Emig, R.L. Jaffe, in *Casimir Physics*, ed. by D. Dalvit, P. Milonni, D. Roberts, F. da Rosa. Lecture Notes in Physics, vol 834 (Springer, Berlin, 2011), pp. 129–174
21. D.J. Griffiths, *Introduction to Electrodynamics*, vol. 3 (Prentice Hall, Upper Saddle River, NJ, 1999)
22. A.D. McLachlan, Mol. Phys. **7**(4), 381 (1964)
23. C. Raabe, D.-G. Welsch, Phys. Rev. A **80**, 067801 (2009)
24. I. Brevik, S.A. Ellingsen, Phys. Rev. A **79**, 027801 (2009)
25. C. Raabe, D.-G. Welsch, Phys. Rev. A **71**, 013814 (2005)
26. E.I. Kats, Zh. Eksp. Teor. Fiz. **73**, 774 (1977)
27. A. Lambrecht, A.C. Duran, R. Guerout, S. Reynaud, in *Casimir Physics*, ed. by D. Dalvit, P. Milonni, D. Roberts, F. da Rosa. Lecture Notes in Physics, vol 834 (Springer, Berlin, 2011), pp. 97–127

# Chapter 3
# The Quantum Nature of the Casimir Force

*Natura abhorret vacuum (Translated: Nature abhors a vacuum).*

Franois Rabelais, *Gargantua and Pantagruel*

## 3.1 Diverging Accounts of the Casimir Effect

We have been considering the Casimir Effect from a number of different theoretical perspectives. In the light of what we have discussed, and before proceeding any further, it is worth taking the time to ask a simple question: what, then, are we actually talking about? Certainly, the Casimir Effect is an empirically verified phenomenon involving attractive (or repulsive) forces between macroscopic objects that persists even at zero temperature in a vacuum [1, 2].

However, explanations of the phenomenon are not uniformly consistent among theorists. The Casimir force has been described, on the one hand, as an effect resulting from the alteration, by the boundaries, of the zero-point electromagnetic energy [3]. On this account, the force is a property of the vacuum and "clear evidence for the existence of vacuum fluctuations" [4]. On the other hand, the Casimir Effect has also been described as a "force [that] originates in the forces between charged particles" that can be "computed without reference to zero point energies". According to this alternative account, "The Casimir force is simply the (relativistic, retarded) Van der Waals force between the metal plates" and the phenomenon offers "no evidence that the zero-point energies are real" [5].

This is rather unsatisfactory [6–8]. We should like to be able to say something clearly (albeit provisionally[1]) about what this phenomenon subsists in, and what it may imply about the nature of physical reality. However, to adopt a more metaphysical parlance, these popular accounts of the Casimir Effect appear to invoke different

[1] Inevitably, this is an interim position. We know that a better theory will eventually be required because of the deep problems in reconciling quantum field theory and gravity.



W.M.R. Simpson, *Surprises in Theoretical Casimir Physics*, Springer Theses,
DOI 10.1007/978-3-319-09315-4_3

ontologies in which a certain metaphysical priority is exchanged between the matter and the fields.

In this chapter, our aim is to side-step the technical details of Casimir physics and reconsider the basic ideas, with the hope of achieving some conceptual clarity. In so doing it will become apparent that these inconsistent interpretations are grounded in theories that fail to offer a consistently quantum-mechanical description of the interaction of the field with matter. There are two sides to this quantum-mechanical coin, but neither appears to be weighted. In the author's opinion, the proper locus for interpreting the Casimir Effect is the theory of macroscopic quantum electrodynamics, in which the necessary quantization of the electromagnetic field and its coupling to bulk materials receives a canonical and consistently quantum-mechanical treatment.

## 3.2 Three Theories of the Casimir Force

### *3.2.1 Casimir's Theory and the Quantum Vacuum*

#### 3.2.1.1 Theoretical Context

We shall focus our efforts on Casimir's classic thought-experiment. In the standard account of the Casimir Effect, the predicted force occurs between a pair of neutral, parallel conducting plates, separated by a distance $d$, in vacuum at zero temperature.[2] The interaction arises due to a disturbance of the vacuum state of the electromagnetic field (in which there are no real photons between the plates) [1]. This is a quantum effect, as classical electrodynamics does not predict a force at zero temperature.

The prescribed procedure may be summarised as follows [3]: take the infinite vacuum energy of the quantized electromagnetic field, with Dirichlet boundary conditions imposed on the field modes,

$$E = \frac{1}{2} \sum \hbar\omega, \tag{3.2.1}$$

and subtract from it the infinite vacuum energy in free Minkowski space (or with the boundaries infinitely separated), $E_\infty$, having first regularized both quantities $E \to E(\xi)$, $E_\infty \to E_\infty(\xi)$ so that the subtraction procedure is well-defined. Once the difference between the two energies has been computed, the regularization is removed, $\xi \to 0$, and the result that remains is finite:

$$E_{Casimir} = \lim_{\xi \to 0} \left[ E(\xi) - E_\infty(\xi) \right]. \tag{3.2.2}$$

[2] See Chap. 1 for a discussion of the original Casimir Effect.

This is the renormalised Casimir energy, from which we can derive the mechanical force exerted on two parallel plates. For Casimir's case, in which the mirrors are perfectly reflective for all frequencies, we find the pressure force

$$P = -\frac{\hbar c \pi^2}{240 d^4}. \tag{3.2.3}$$

The astute reader will rightly object that our own procedure for extracting Casimir's result in Chap. 1 differed slightly from this recipe, and indeed nobody follows the standard prescription precisely for the case of the electromagnetic field, though it has been pedagogically applied to a 1d scalar field [3] where the calculation is somewhat simpler. As we have seen, if we apply a frequency cutoff term $exp(-\xi\omega/c)$ as the regulariser, we discover an additional divergent term that is not removed by subtracting the so-called background energy [9], which appears to correspond to waves running parallel to the plates.[3] This extra term has to be discarded also in order to state a finite electromagnetic Casimir energy. Typically it disappears in the course of applying the Euler-MacLaurin formula (e.g. [7, 11, 12]), or as a result of applying some other mathematical trick where the physical meaning is difficult to discern. Suffice it to say that the simple picture of taking the difference between two electromagnetic field energies can be somewhat misleading.

#### 3.2.1.2 Physical Interpretation

Nevertheless, considered on the basis of an energy mode summation, as employed by Casimir [1], it seems the quantised electromagnetic field in its ground-state with 'external boundary conditions' is sufficient to determine a force—an almost matter-free prescription for obtaining the phenomenon in which the interacting bodies become simply topological features of the space [7]. Casimir's formula, depending solely upon the constants $\hbar$ and $c$ and the distance $d$ between the plates, serves to consolidate this impression [5].

But this interpretation is naive. The vacuum energy, as we have observed, is infinite, and in addition to imposing boundary conditions on the field we apply some form of regularisation to tame the mode summation and permit the subtraction (or extraction) of diverging terms. Although the various mathematical techniques employed to do this often obscure the fact, it is in the procedure of regularisation that some of the properties of matter—in particular, its dispersive behaviour—are permitted to leak into the calculation, albeit rather crudely [9]. Significantly, it is not possible to extract anything meaningful (or measureable) about the Casimir force until they are permitted to do so. Furthermore, when we relax the highly artificial boundary condition of perfect mirrors, as we must in order to predict the Casimir Effect in real materials, we are forced to sum contributions to the Casimir energy over a dispersive material response across the whole mode spectrum, substantially

[3] This additional term appears as the second term in Eq. (1.1.21).

modifying the predicted force. To do this kind of calculation, however, we must abandon the mode summation and adopt a more sophisticated apparatus, like Lifshitz theory. Casimir's result can still be recovered, but only as a limiting case [11].

## 3.2.2 *Lifshitz Theory and Stochastic Fluctuations*

### 3.2.2.1 Theoretical Context

Lifshitz theory, which we discussed in Chap. 2, has proven an important benchmark for the prediction of Casimir forces in more realistic cases, enjoying significant experimental verification [13, 14]. In the context of Lifshitz theory, the Casimir Effect is a result of fluctuating current densities in the two plates [15–17]. A force arises from the interaction of the currents through the electromagnetic field that they generate in the cavity. The plates are now treated more realistically as dielectric with frequency-dependent permittivities and permeabilities, and this substantially affects both the size (and, in some cases, the nature[4]) of the predicted force.

The formalism is written in terms of the electromagnetic Green function, which describes the field produced by sources of current within the system (2.2.30), including the stress tensor from which the force is derived (2.3.75, 2.3.76). The stress tensor, however, like the zero-point energy, contains a divergent contribution that must also be regularised.[5] Typically this is achieved by subtracting a stress $\sigma_0$ calculated using an auxiliary Green function associated with an infinite homogeneous medium [11, 17–20], and computing the physical stress in the limit of the point of measurement approaching a point source:

$$\sigma_{Casimir} = \lim_{\mathbf{r}' \to \mathbf{r}} \left[ \sigma(\mathbf{r}, \mathbf{r}') - \sigma_0(\mathbf{r}, \mathbf{r}') \right]. \tag{3.2.4}$$

One can then determine a finite stress tensor for the system that depends on the dielectric functions of the material at imaginary frequencies (quantities obtained from the dielectric properties for real frequencies by Hilbert transformation). Only then can the force be derived. Both Casimir's and Lifshitz' regularizations give identical results in the limiting case of a cavity sandwiched between perfectly reflecting mirrors (2.4.31).

---

[4] Lifshitz theory predicts repulsive Casimir forces, under certain circumstances [35].

[5] As we shall see, additional divergences in the stress appear in the generalisation to inhomogeneous media (where the optical properties vary continuously along at least one spatial axis). In this case, the regularisation cannot remove the infinities [9, 10, 36, 37].

#### 3.2.2.2 Physical Interpretation

An incautious reading of Lifshitz theory might suggest that the role of the vacuum has been successfully banished from the Casimir Effect. Ontologically, the conditions seem to involve merely the material in the plates and a stochastic source of fluctuations within the material. An electromagnetic field results from the fluctuating currents in the plates, producing a mechanical stress in the same pieces of material that generated it. There is no Hamiltonian in the original formulation of Lifshitz theory [15, 16] and there are no quantized fields. There is therefore no ground-state of the electromagnetic field. To some, this does not even appear to be a quantum-mechanical theory at all [18, 21–23].

But the fluctuations in the material that persist even at zero temperature are not a classical phenomenon; they are inserted 'by hand' using Rytov's correlation function. This can be derived from statistical physics, or from fluctuation-dissipation theorem,[6] and affords the average electromagnetic field that would be present at finite (or zero) temperature [24].

Lifshitz theory is arguably uncommitted to the particulars of a quantum theory of light in material, however, as opposed to a merely stochastic theory of the phenomena, being based rather on the principles of thermodynamics and statistical physics [18, 21]. It embodies at best a minimal treatment of the quantum mechanics that is phenomenologically driven. It is therefore ontologically ambiguous about the role of the vacuum.[7] A clearer interpretation of the underlying physics cannot be achieved without the vantage point of a more quantum-mechanically consistent position.

### 3.2.3 *Macroscopic QED and the Polariton Field*

#### 3.2.3.1 Theoretical Context

A recent and more sophisticated formulation of macroscopic quantum electrodynamics than the kind we adopted in Chap. 2 offers a canonical quantum-mechanical treatment of the interaction of light with real materials, without the detailed reference to the microscopic material structure that must defeat any complete treatment of such systems, and without sacrificing quantum-mechanical consistency along with more phenomenologically driven approaches [21]. Significantly, this form of macro-QED is the only *canonical* method that reproduces and justifies the general Lifshitz theory result for the stress tensor and determines the Casimir energy density *inside* a

---

[6] See Sect. 2.2.

[7] Some argue for the consistency of Lifshitz theory with Casimir's approach. Bordag writes: "Lifshitz considered the fluctuations in the medium as source. In the modern understanding, these two are equivalent. However, the discussion about two ways continues until present time" [28]. Schwinger, on the other hand, seems to exploit the ambiguity of Lifshitz' theory with his 'source theory', replacing the fluctuations of the vacuum with source fields in the plates, with the intention of removing any references to a vacuum state with non-zero physical properties [27].

medium [16], resolving a long-standing dispute over the form that these expressions should take in this context [18, 20, 25, 26]. It applies with full generality to arbitrary magnetodielectrics, taking full account of the phenomena of dispersion and dissipation. First, in distinction to the shortcuts taken in Chap. 2, an action is formulated in terms of the dynamical variables $\{\phi, \mathbf{A}\}$, the scalar and vector potentials of the fields, and $\{\mathbf{X}_\omega, \mathbf{Y}_\omega\}$, a pair of oscillator fields incorporating the dissipative behaviour of the material:

$$S[\phi, \mathbf{A}, \mathbf{X}_\omega, \mathbf{Y}_\omega] = S_{em}[\phi, \mathbf{A}] + S_X[\mathbf{X}_\omega] + S_Y[\mathbf{Y}_\omega] + S_{int}[\phi, \mathbf{A}, \mathbf{X}_\omega, \mathbf{Y}_\omega], \quad (3.2.5)$$

where $S_{em}$ is the free electromagnetic action, $S_X$ and $S_Y$ are the actions for the free reservoir oscillators, and $S_{int}$ is the interaction part of the action, coupling the electromagnetic fields to the field reservoirs of the material. Maxwell's equations can be recovered from this action, and canonical quantisation proceeds straightforwardly. As before (2.1.40), a diagonalised Hamiltonian is achieved,[8]

$$\hat{H} = \frac{1}{2} \sum_{\lambda=e,m} \int d^3\mathbf{r} \int_0^\infty d\omega\, \hbar\omega \left( \hat{\mathbf{f}}_\lambda^\dagger(\mathbf{r}, \omega) \cdot \hat{\mathbf{f}}_\lambda(\mathbf{r}, \omega) + \hat{\mathbf{f}}_\lambda(\mathbf{r}, \omega) \cdot \hat{\mathbf{f}}_\lambda^\dagger(\mathbf{r}, \omega) \right), \quad (3.2.6)$$

where the eigenmodes are bosonic creation and annihilation operators obeying commutation relations

$$\left[ \hat{f}_{\lambda i}(\mathbf{r}, \omega), \hat{f}_{\lambda' j}^\dagger(\mathbf{r}', \omega') \right] = \delta_{ij}\delta_{\lambda\lambda'}\delta(\omega - \omega')\,\delta(\mathbf{r} - \mathbf{r}'), \quad (3.2.7)$$

$$\left[ \hat{f}_{\lambda i}(\mathbf{r}, \omega), \hat{f}_{\lambda' j}(\mathbf{r}', \omega') \right] = 0. \quad (3.2.8)$$

Charge and density operators for the material in the plates, as well as operators for the electromagnetic field, are then expressed in terms of the creation and annihilation operators of the system.

Casimir forces are caused by the stress-energy of the electromagnetic fields in a state of thermal equilibrium. We require the eigenmodes of the system to be in a thermal mixed quantum state. To determine the Casimir force, we consider the ground-state of the system and compute the electromagnetic part of the energy density or stress tensor [18], where the complete stress-energy-momentum tensor of the system is obtained self-consistently from the application of Noether's theorem to the original action. The Casimir stress tensor is recovered as the expectation value of the electromagnetic component in thermal equilibrium. At zero-temperature, we recover the zero-point Casimir stress, which has the same form as the more general result for

[8] The zero-point term is suppressed in the statement of the diagonalised Hamiltonian in the original paper [21], but a zero-point energy is present nonetheless.

the stress tensor in Lifshitz theory, and from which the Casimir forces in the system can finally be determined [16], once the stress tensor has been regularised.[9]

#### 3.2.3.2 Physical Interpretation

The canonical theory of macroscopic quantum electrodynamics—at least, at present—seems to offer the richest and most general basis for interpreting Casimir phenomena. In macro-QED the materials and the fields are placed on a more equal footing. Physics necessitates the quantization of the fields, and a consistent account of quantised light in media demands a quantum representation of the relevant properties of the material. In the Hamiltonian for macro-QED, both aspects of the system are quantised and coupled. The quantum of this system—that is, its irreducible unit of excitation—is a type of polariton.

This in turn means that the matter is coupled to the vacuum state of the fields. Under the condition of thermal equilibrium,[10] it follows that the ground-state of the total system, including both the matter and the fields, is endowed with physical properties. To see this, consider the following: the fluctuating currents in one plate only interact with the currents in the other if they *communicate* with them, and they communicate through the electromagnetic field. At zero temperature, there are no photons between the plates, on pain of violating thermodynamic equilibrium; the electromagnetic field is therefore in its ground state. At zero temperature, therefore, the currents in the plates can only communicate through the zero-point radiation.

There is another sense in which macro-QED places the matter and the fields on an equal footing: the Lagrangian formulation that underlies the action, in which the fields, the material and their interaction are posited, is acausal in this respect: we could view the matter as producing the fields, or we could view the fields as inducing the currents in the matter; the actual physics of the phenomenon does not prioritise either.[11]

## 3.3 Three Ontologies of the Casimir Effect

### *3.3.1 Semi-classical Ontologies*

Broadly speaking, it is possible to characterise the polarisation of opinion (or preference) concerning the Casimir Effect into two ontologically distinct positions in which a certain metaphysical priority is exchanged:

---

[9] See Sect. 2.3.3.

[10] Lifshitz' stress tensor, commonly used for calculating Casimir forces in realistic systems, is in fact derived under the condition of thermal equilibrim [15, 16, 25].

[11] The dynamics are obtained by extremising the action.

(I) a *vacuum-field* interpretation: there exists an electromagnetic quantum vacuum field, whose properties are modified by material (or topological) constraints, which gives rise to forces between material bodies.
(II) a *fluctuating material* interpretation: there exist distributions of spontaneously polarised material, whose quantum fluctuations give rise to forces between them through the retarded electromagnetic field.

The first account touts the existence of a fluctuating electromagnetic quantum field in its ground-state—the quantum vacuum, a state that is void of particles or quanta, but has the property of an energy available for doing work (for an example of this view, see [3]). This property of the field is modified by the imposition of material (or topological) constraints. In the case of the two parallel plates, the energy of the vacuum is reduced (made more negative) by the motion of the plates towards each other. The attractive Casimir force that results is thus a consequence of the zero-point energy of the vacuum; that is, the energy associated with the fluctuations of the vacuum.

On the second account, it is claimed that one is not required to invoke electromagnetic quantum fluctuations of the vacuum to explain Casimir phenomena, and the Casimir force is essentially reinterpreted as a giant van der Waals effect (for examples of this approach, see [5, 27]). The material in the plates (as opposed to any field between the plates) is subject to quantum fluctuations. These spontaneous disturbances produce field-generating currents within the plates, which interact through the retarded electromagnetic field they have created. These interactions result in a force between the plates.

The seasoned theorist may put it down to a matter of personal taste as to which of the two approaches is preferable, arguing that either position is empirically adequate [28]. The ingeniously contrived path-integral scattering method developed in [29], for instance, obtains two equivalent representations of the Casimir energy, one in terms of fluctuating fields and the other in terms of fluctuating charges.

However, the theories in which these approaches are typically grounded do not enjoy both the consistency and the generality of macro-QED. For example, in addition to offering a canonical quantum mechanical theory of light in media and determining the general stress tensor, macro-QED is being fruitfully applied to the problem of electromagnetic effects resulting from the motion of dielectrics, including the hotly disputed problem of quantum friction, for which there is now a clear and sophisticated answer taking full account of the phenomena of dispersion and dissipation [30, 31]. Moreover, accounts (I) and (II) are ontologically divided insofar as the first requires a quantum vacuum state with physical properties and the second does not. The van der Waals interpretation does not assign any physical properties to the vacuum. Importantly, neither of these interpretations is grounded in both a general and consistently quantum-mechanical description of the interaction of the field with matter; typically, the quantum mechanical-treatment of the problem falls unevenly on one aspect or the other.

### 3.3.2 A Dual-Aspect Quantum Ontology

Let us consider instead this third option:

(III) a *dual-aspect quantum* interpretation : there exists a vacuum state of the coupled system of matter and fields, which determines the ground-state properties of the electromagnetic field, giving rise to a force.

On this interpretation, the Casimir force is fundamentally a property of the coupled system of the matter and the fields, in which the interaction between the plates is mediated by the zero-point fields [18, 21]. This interpretation affirms and denies different tenets of interpretations (I) and (II):

First, in common with (I) rather than (II), there is a vacuum state in (III) which has a physical energy and a role to play in the Casimir Effect. That is, despite the absence of photons between the plates, and thermal vibrations within the plates, the walls of the cavity will still experience an attractive force. Contra (I), but in consort with (II), however, the Casimir Effect does not warrant the assignment of physical meaning to the energy of the vacuum state of the field as a *ding-an-sich*,[12] or totting up its modes in an enormous contribution to the cosmological constant (it is typically cut off at the Planck scale) [32, 33]. The Casimir Effect offers no justification for quantizing the plane waves of an infinite homogeneous space (which presupposes no coupling to matter) and reifying the 'zero-point energy' obtained. There is no force in that case, nor anything for a force to act on. In the quantisation of light coupled to matter, however, the modes of the field are no longer characterised by plane waves. In other words, it is in this state of interaction that we determine the Casimir energy and measure a Casimir force. Regularisation amounts to drawing a perimeter around this interaction.

This is enough clarification for our purposes. We have described the requisite ontology of the Casimir Effect as 'dual-aspect', an appelation that is intended to be sufficiently generous to encompass more detailed accounts within the constraints we have discussed. In some sense, the electromagnetic and material aspects of the system are simultaneously present in the vacuum state. However, it is not without the additional structure involved in the details of their interaction that they contribute any actual properties to the system that make contact with observable reality. This irreducible character of the system, in which the 'whole' is more than the sum of its 'parts', is formally represented in the Hamiltonian (via the action) through the addition of an interaction term. In seeking a more detailed account, perhaps we may take our cue from Heisenberg, who saw the wave function as neither the description of any actual state of affairs, nor merely a convenient calculating device, but as referring to a kind of potentiality [34], and would presumably have described the vacuum similarly.[13] Perhaps for others, the language of emergence may prove the

[12] That is, a *thing in itself.*

[13] Saunders' observation is sapiential. He writes: 'on every other of the major schools of thought on the interpretation of quantum mechanics [besides stochastic hidden variable theories]... there is no

more useful in relating the different aspects of this problem. These are conceptual problems that call for further discussion elsewhere.

## 3.4 Summary Remarks

In a general and consistently quantum-mechanical theory of light in media, the Casimir Effect may be properly described as a force arising out of the ground-state properties of a polariton field—a coupled, quantised system of dielectric material and electromagnetic fields. In its ground state, the system cannot be separated into material or electromagnetic quanta, since none have been excited. Nevertheless, a Casimir force is predicted between the plates.

In interpreting Casimir's theory, however, the metaphysical emphasis seems to have fallen either on the electromagnetic field or the fluctuating material in the plates. On the first interpretation (I), a universal vacuum field is postulated, in which the ontological role of matter is (sometimes minimally) acknowledged in the form of boundary conditions and (sometimes unconsciously) in the regularisation process. On the second interpretation (II), the bulk material in the plates is prioritised, in which the fluctuating currents generate a field between the cavity that attracts the plates together. The role of the vacuum is void.

However, neither of these interpretations is entirely adequate, and this may lead to false predictions. For example, in opposition to (I), there is no reason to suppose that the energy of the vacuum state in the absence of any coupling is real, or that the Casimir Effect validates the huge contribution to the cosmological constant that this hypostatisation entails. The Casimir force, ironically, should not be seen through the lens of Casimir's calculation, which is itself without physical application, being at best the limiting case of a more complicated, more abstract, and more consistently quantum-mechanical theory, in which the effects of matter and light are inextricably intertwined.

## References

1. H.B.G Casimir, Koninkl. Ned. Akad. Wetenschap., **51**, 793 (1948)
2. S.K. Lamoreaux, Phys. Rev. Lett. **78**, 5 (1997)
3. M. Bordag, U. Mohideen, V.M. Mostepanenko, Phys. Rep. **353**, 1–205 (2001)
4. S.M. Carroll, Living Rev. Rel. **4**, 1 (2001)
5. R.L. Jaffe, Phys. Rev. D **72**(021301(R)) (2005)

(Footnote 13 continued)
reason to suppose that the observed properties of the vacuum, when correlations are set up between fields in vacuo and macroscopic systems, are present in the absence of such correlations' [6]. Boi notes the possibility of taking 'the vacuum as a kind of pre-substance, an underlying substratum having a potential substantiality. It can... become physical reality if various other properties are injected into it' [7].

6. S. Saunders, *Ontological Aspects of Quantum Field Theory* (World Scientific, London, 2002)
7. L. Boi, *The Quantum Vacuum* (The Johns Hopkins University Press, Baltimore, 2011)
8. S. Saunders, H.R. Brown (eds.), *The Philosophy of Vacuum* (Clarendon Press, Oxford, 1991)
9. S.A.R. Horsley, W.M.R. Simpson, Phys. Rev. A **88**(013833) (2013)
10. W.M.R. Simpson, S.A.R. Horsley, U. Leonhardt, Phys. Rev. A **87**, 043806 (2013)
11. U. Leonhardt, *Essential Quantum Optics* (Cambridge University Press, Cambridge, 2010)
12. P.W. Milonni, *The Quantum Vacuum* (Academic Press, New York, 1994)
13. J.N. Munday, F. Capasso, V.A. Parsegian, Nature **457**, 170 (2009)
14. A.W. Rodriguez, F. Capasso, S.G. Johnson, Nat. Photonics **5**, 211 (2011)
15. E.M. Lifshitz, Zh. Eksp. Teor. Fiz. **29**, 94 (1955)
16. I.E. Dzyaloshinskii, E.M. Lifshitz, L.P. Pitaevskii, Adv. Phys. **10**, 165 (1961)
17. E.M. Lifshitz, L.P. Pitaevskii, *Statistical Physics (Part 2)*. (Butterworth-Heinemann, Oxford, 2003)
18. T.G. Philbin, New. J. Phys. **13**, 063026 (2011)
19. T.G. Philbin, C. Xiong, U. Leonhardt, Ann. Phys. **325**, 579 (2009)
20. L.P. Pitaevskii, In *Casimir physics*, Vol. 834, ed. by Diego Dalvit, Peter Milonni, David Roberts, Felipe da Rosa. Lecture Notes in Physics. (Springer, Berlin, 2011), pp 23–37
21. T.G. Philbin, New J. Phys. **12**, 123008 (2010)
22. L. Knöll, S. Scheel, D.-G. Welsch, *Coherence and Statistics of Photons and Atoms* (Wiley, New York, 2001)
23. G. Barton, New J. Phys. **12**(113045) (2010)
24. F.S.S Rosa, D.A.R Dalvit, P.W. Milonni, Phys. Rev. A **81**, 033812 (2010)
25. L.P. Pitaevskii, Phys. Rev. A **73**, 047801 (2006)
26. C. Raabe, D.-G. Welsch, Phys. Rev. A **71**, 013814 (2005)
27. J. Schwinger, L.L. DeRaad, K.A. Milton, ANNALS Ann. Phys. **115**, 1–23 (1978)
28. M. Bordag, Phys. Rev. D, **85**(025005) (2012)
29. S.J. Rahi, T. Emig, N. Graham, R.L. Jaffe, M. Kardar, Phys. Rev. D **80**, 085021 (2009)
30. S.A.R. Horsley, Phys. Rev. A **84**, 063822 (2011)
31. S.A.R. Horsley, Phys. Rev. A **86**, 023830 (2012)
32. Brandenberger, H. Robert, arXiv:hep-th/0210165, 2002
33. S. Weinberg, Rev. Mod. Phys. **61**, 1–21 (1981)
34. W. Heisenberg, *Physics and Philosophy: The Revolution in Modern Science* (Penguin Classics, New York, 2000)
35. F. Capasso, J.N. Munday, *Casimir Physics: Lecture Notes in Physics*, vol 834 (Springer, Berlin, 2011)
36. U. Leonhardt, W.M.R. Simpson, Phys. Rev. D **84**, 081701 (2011)
37. W.M.R. Simpson, Casimir force in a compressive transformation medium. Phys. Rev. A **88**(6), 063852 (2013)

# Part II
# Surprises in Casimir Theory

# Chapter 4
# The Cut-off Independence of the Casimir Energy

*"Begin at the beginning," the King said, gravely, "and go on till you come to an end; then stop."*

Lewis Caroll, *Alice in Wonderland*

## 4.1 Beyond Homogeneous Media

In Chap. 1, we considered the ground state energy of the electromagnetic field in a piston geometry (see Fig. 1.2) for the idealised case where the piston and the walls of the chamber are taken as perfect mirrors. A moveable mirror is positioned at $x = a$. When the chamber is empty we can recover straightforwardly Casimir's well-known expression for the Casimir force (1.1.23). We can also determine the Casimir force for the case of a cavity filled with a homogeneous medium.[1] In cases such as these, the Casimir-energy of the system can be regularised and is cut-off-independent, and the Casimir pressure on the piston is finite and independent of the small scale physics of the media that compose the mirrors.

However, suppose the chamber of the piston were to be filled with a dielectric material, in which the refractive index of the medium varies continuously along at least one axis. Whilst Casimir forces between macroscopic bodies have been calculated for a variety of systems and geometries [1, 2], it is difficult to find examples in which the optical properties of the interacting media have not been idealised as perfectly homogeneous. The purpose of the following discussion is to consider the Casimir energy for a system incorporating a non-dispersive inhomogeneous medium.[2] We find that the force on the mirror can be stated exactly when computed from a simple summation over the field modes.

---

[1] See Sect. 2.4.

[2] This chapter develops and responds to the argument published in [3], which found a cutoff-dependence in the Casimir energy when the mirrors are separated by an inhomogeneous medium. However, this derivation contains an error that substantially affects the results of this paper: equation (15), describing the electromagnetic field for one of two polarisations, is incorrectly normalised for modes $m = 0$. I am grateful to Fanglin Bao for pointing this out.

 
W.M.R. Simpson, *Surprises in Theoretical Casimir Physics*, Springer Theses,
DOI 10.1007/978-3-319-09315-4_4

## 4.2 An Inhomogeneous Casimir Piston

In our earlier derivation of the Casimir Effect, the procedure for rendering the ground-state energy (1.1.1) convergent[3] yielded terms that, if dependent on $a$, remained finite or vanished in the limit $\xi \to 0$. That is to say, the Casimir force in such cases is cut-off-independent. However, it is not obvious whether this fortuitous situation occurs for fundamental reasons. Suppose we take the same cavity with an *inhomogeneous* medium within the chamber (for example, Fig. 4.1), in which we assume a permeability and permittivity given by

$$\mu(x) = \mu_0,$$
$$\epsilon(x) = \epsilon_0 \left[1 + \delta\epsilon(x)\right]. \tag{4.2.1}$$

Are the divergent terms in the ground-state energy still independent of $a$? We will imagine this medium as a rigid body, so that we can neglect any energy associated with deformation, and attempt to find the behaviour of the force necessary to hold the mirror fixed at $a$ by computing the mechanical energy of the system. Let us begin by considering the particular case illustrated in Fig. 4.1,

$$f(x) = \frac{1}{2}\left[1 - \cos(2\pi x/L)\right], \tag{4.2.2}$$

where the inhomogeneity is written as $\delta\epsilon(x) = \delta\alpha f(x)$, and $\delta\alpha$ is a small parameter describing the size of the perturbation. We emphasise that we are considering some arbitrary fixed value of $L$. The quantity $L$ only occurs in this expression in order to make the permittivity equal to that of free space at the edges of the cavity.

### 4.2.1 The Modified Casimir Energy

In general it is complicated to find the eigenfrequencies of the electromagnetic field in an inhomogeneous medium. Even with closed form expressions for these frequencies, performing the energy mode-summation (1.1.13) is not straightforward. To proceed we calculate the rate of change of the energy of the system as an inhomogeneous index profile is introduced into the cavity. For our purposes, this perturbation need only be very small. The inhomogeneity (4.2.2) induces a corresponding shift in the eigenfrequencies,

$$\omega_{m,\lambda} = \omega^{(0)}_{m,\lambda} + \delta\alpha\, \omega^{(1)}_{m,\lambda}, \tag{4.2.3}$$

where $m$ labels the spatial dependence of the mode, and $\lambda \in \{1, 2\}$ the polarisation. The first term in (4.2.3) denotes the original frequency corresponding to the case of vacuum, and the second a small perturbation produced by the presence of the

[3] See Sect. 1.1.2.

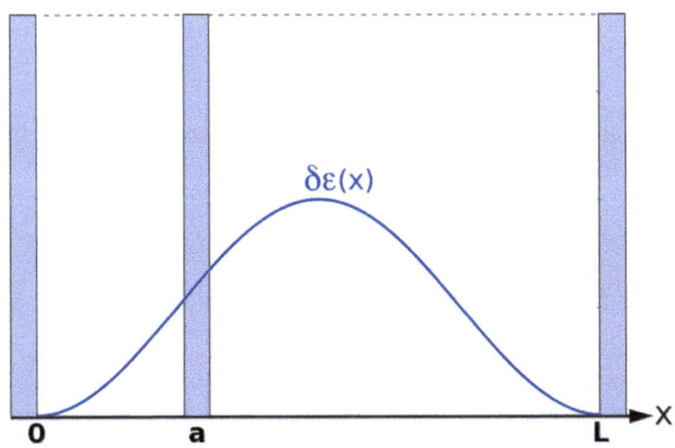


**Fig. 4.1** As in Fig. 1.2, we consider a perfectly reflecting rectangular chamber of length $L$ and cross sectional area, $A$. Within the chamber is a mirror, positioned at $x = a$, and surrounded by an inhomogeneous dielectric with $\epsilon$ and $\mu$ given by (4.2.1). We seek the dependence of the energy of this system on the parameter $a$

medium. Assuming the same regularisation as before (1.1.14), the change of the energy as $\delta\epsilon(x)$ is increased from zero is given by

$$\frac{\tilde{E}+\delta\tilde{E}}{A} = \frac{\hbar}{4\pi}\sum_{m,\lambda}\int_0^\infty k_\parallel dk_\parallel \left(\omega_{m,\lambda}^{L(0)} + \delta\alpha\,\omega_{m,\lambda}^{L(1)}\right) \times e^{-\xi\left(\omega_{m,\lambda}^{L(0)}+\delta\alpha\,\omega_{m,\lambda}^{L(1)}\right)} + \{\mathrm{L} \rightarrow \mathrm{R}\}. \tag{4.2.4}$$

After a first-order expansion of the second term of the exponent, the rate of change of the energy, as the strength of the inhomogeneity $\delta\alpha$ is gradually increased, may be given by the expression

$$\frac{1}{A}\frac{\delta\tilde{E}}{\delta\alpha} = \frac{\hbar}{4\pi}\frac{\partial}{\partial\xi}\sum_{m,\lambda}\xi\int_0^\infty k_\parallel dk_\parallel\left[\omega_{m,\lambda}^{L(1)} e^{-\xi\omega_{m,\lambda}^{L(0)}(k_\parallel)/c} + \omega_{m,\lambda}^{R(1)} e^{-\xi\omega_{m,\lambda}^{R(0)}(k_\parallel)/c}\right]. \tag{4.2.5}$$

### *4.2.2 The Modified Eigenfrequencies*

To calculate (4.2.5) we must now determine the first order shifts in the eigenfrequencies. The situation is slightly more involved than before, due to the fact that the two polarisations do not behave degenerately in the medium. To calculate the change in the eigenfrequencies of the cavity we use first order perturbation theory. Within the cavity, the electromagnetic field obeys,

$$\nabla\times\nabla\times\boldsymbol{E}_{m,\lambda} - \frac{\omega_{m,\lambda}^2}{c^2}[1+\delta\epsilon(x)]\boldsymbol{E}_{m,\lambda} = 0, \tag{4.2.6}$$

where $m$ labels the spatial dependence of the mode, and $\lambda \in \{1, 2\}$ the polarisation. When $\delta\epsilon = 0$, the modes in the region $x \in [0, a]$ are given by

$$\boldsymbol{E}_{m,1}^{(0)} = \hat{\boldsymbol{x}}\times\hat{\boldsymbol{k}}_{\parallel}\sqrt{\frac{2}{aA}}\sin(k_x x)e^{i\boldsymbol{k}_{\parallel}\cdot\boldsymbol{x}}, \tag{4.2.7a}$$

$$\boldsymbol{E}_{m,2}^{(0)} = \sqrt{\frac{(2-\delta_{m0})/aA}{k_{\parallel}^2+(m\pi/a)^2}}\left[k_{\parallel}\hat{\boldsymbol{x}}\cos(k_x x) - i\hat{\boldsymbol{k}}_{\parallel}\left(\frac{m\pi}{a}\right)\sin(k_x x)\right]e^{i\boldsymbol{k}_{\parallel}\cdot\boldsymbol{x}}, \tag{4.2.7b}$$

where $k_x = m\pi/a$ and $\mathbf{k}_{\parallel} = k_y\hat{\mathbf{y}} + k_z\hat{\mathbf{z}}$. We assume the limit of $L_y/a, L_z/a \to \infty$. The modes in the region $x \in [a, L]$ are also given by (4.2.7a, 4.2.7b) after substituting $a \to L - a$ and $x \to x - a$. All modes are normalised over the volume $V$ of each region of the chamber; the Kronecker delta function in 4.2.7b ensures that the volume integral for $m = 0$ is also unity. Writing the perturbed field as

$$\boldsymbol{E}_{m,\lambda} = \boldsymbol{E}_{m,\lambda}^{(0)} + \delta\alpha\boldsymbol{E}_{m,\lambda}^{(1)}, \tag{4.2.8}$$

and inserting (4.2.3) and (4.2.8) into Eq. (4.2.6), the wave equation becomes

$$\begin{aligned}&\nabla\times\nabla\times\mathbf{E}_{m,\lambda}^{(0)} + \delta\alpha\,\nabla\times\nabla\times\mathbf{E}_{m,\lambda}^{(1)}\\ &-\frac{1}{c^2}\left(\omega_{m,\lambda}^{(0)2} + \omega_{m,\lambda}^{(1)2}\delta\alpha^2 + 2\delta\alpha\omega_{m,\lambda}^{(0)}\omega_{m,\lambda}^{(1)}\right)(1+\delta\alpha\, f(x))\left(\mathbf{E}_{m,\lambda}^{(0)} + \delta\alpha\mathbf{E}_{m,\lambda}^{(1)}\right) = 0.\end{aligned} \tag{4.2.9}$$

This modified wave equation can be simplified using a few elementary tricks. To begin with, some of the terms must cancel each other. We know that, for empty space

$$\nabla\times\nabla\times\mathbf{E}_{m,\lambda}^{(0)} - \frac{\omega_{m,\lambda}^{(0)2}}{c^2}\mathbf{E}_{m,\lambda}^{(0)} = 0. \tag{4.2.10}$$

Also, the perturbation is small. Retaining only terms up to first order and dividing by $\delta\alpha$, we find that, to first order in $\delta\alpha$

$$\nabla\times\nabla\times\boldsymbol{E}_{m,\lambda}^{(1)} - \frac{1}{c^2}\left[2\omega_{m,\lambda}^{(1)}\omega_{m,\lambda}^{(0)}\boldsymbol{E}_{m,\lambda}^{(0)} + \omega_{m,\lambda}^{(0)2}\boldsymbol{E}_{m,\lambda}^{(1)} + \omega_{m,\lambda}^{(0)2}\boldsymbol{E}_{m,\lambda}^{(0)}f(x)\right] = 0. \tag{4.2.11}$$

Exploiting the orthogonality of the modes,

$$\int d^3x\,\mathbf{E}_{m,\lambda}^{(0)}\cdot\mathbf{E}_{p,\lambda'}^{(0)*} = \delta_{m,p}\delta_{\lambda,\lambda'}, \tag{4.2.12}$$

we proceed by multiplying (4.2.11) on the left by $\mathbf{E}^{(0)*}_{p,\lambda'}$ and integrating over $V$ (which could either be the left or right region of Fig. 4.1):

$$\int d^3x\,\mathbf{E}^{(0)*}_{p,\lambda'}\cdot\left(\nabla\times\nabla\times\mathbf{E}^{(1)}_{m,\lambda}\right)-\frac{1}{c^2}\left\{\omega^{(0)2}_{m,\lambda}\int d^3x\,\mathbf{E}^{(0)*}_{p,\lambda'}\cdot\mathbf{E}^{(1)}_{m,\lambda}\right.$$
$$+\,2\,\omega^{(0)}_{m,\lambda}\omega^{(1)}_{m,\lambda}\delta_{m,p}\,\delta_{\lambda,\lambda'}$$
$$\left.+\,\omega^{(0)2}_{m,\lambda}\int d^3x\,f(x)\mathbf{E}^{(0)*}_{p,\lambda'}\cdot\mathbf{E}^{(0)}_{m,\lambda}\right\}=0. \tag{4.2.13}$$

We consider the case $p=m$ and $\lambda=\lambda'$:

$$\int d^3x\,\mathbf{E}^{(0)*}_{m,\lambda}\cdot\left(\nabla\times\nabla\times\mathbf{E}^{(1)}_{m,\lambda}\right)-\frac{1}{c^2}\left\{\omega^{(0)2}_{m,\lambda}\int d^3x\,\mathbf{E}^{(0)*}_{m,\lambda}\cdot\mathbf{E}^{(1)}_{m,\lambda}\right.$$
$$+\,2\omega^{(0)}_{m,\lambda}\omega^{(1)}_{m,\lambda}$$
$$\left.+\,\omega^{(0)2}_{m,\lambda}\int d^3x\,f(x)\left|\mathbf{E}^{(0)}_{m,\lambda}\right|^2\right\}=0. \tag{4.2.14}$$

Using integration by parts[4] and requiring that the fields vanish at infinity, we can rewrite the first term as

$$\int d^3x\,\mathbf{E}^{(0)*}_{m,\lambda}.\left(\nabla\times\nabla\times\mathbf{E}^{(1)}_{m,\lambda}\right)=\int d^3x\,\mathbf{E}^{(1)}_{m,\lambda}.\left(\nabla\times\nabla\times\mathbf{E}^{(0)*}_{m,\lambda}\right). \tag{4.2.15}$$

The double-curl on the right-hand side of (4.2.15) can be replaced using the wave equation, changing the term to

$$\frac{\omega^{(0)2}_{m,\lambda}}{c^2}\int d^3x\,\mathbf{E}^{(0)*}_{m,\lambda}.\mathbf{E}^{(1)}_{m,\lambda}. \tag{4.2.16}$$

Hence from (4.2.14) to (4.2.16) it follows that the perturbation in the frequency can be expressed in the simple form

$$\omega^{(1)}_{m,\lambda}=-\frac{1}{2}\omega^{(0)}_{m,\lambda}\int_V\left|\boldsymbol{E}^{(0)}_{m,\lambda}\right|^2 f(x)\,d^3\boldsymbol{x}, \tag{4.2.17}$$

---

[4] The identity is

$$\int_\Omega(\nabla\times\mathbf{u})\cdot\mathbf{v}\,d^3x=-\int_{\partial\Omega}(\mathbf{u}\times\mathbf{n})\cdot\mathbf{v}\,d^2x+\int_\Omega(\nabla\times\mathbf{v})\cdot\mathbf{u}\,d^3x,$$

which is the standard expression for the first order perturbation of the eigenfrequencies of an optical cavity (for examples, see [4, 5]).

### 4.2.3 Frequency Perturbations

Having obtained the general expression, we deduce the perturbations in the eigenfrequencies for our particular example by inserting (4.2.7a, 4.2.7b) into (4.2.17), using the permittivity profile (4.2.2). The piston is symmetrical along the $y$ and $z$ axis, therefore the integral over volume can be replaced by the cross-sectional area $A$, multiplied by an integral over the left or right side of the cavity (i.e. between $x \in [0, a]$ or $x \in [a, L]$).

#### 4.2.3.1 Frequency Perturbation of the First Polarisation

For the first polarisation, $\lambda = 1$:

$$\omega^{(1)}_{m,1} = -\frac{\omega^{(0)}_{m,1}}{2a} \int_0^a \left[\sin^2 (m\pi x/a) \left[1 - \cos (2\pi x/L)\right]\right] \mathrm{d}x. \tag{4.2.18}$$

We note that for $m = 0$ there is no first-order correction. The integral above can be separated and determined straightforwardly. For the first integral:

$$\int_0^a \sin^2 (m\pi x/a)\, \mathrm{d}x = \frac{a}{2}. \tag{4.2.19}$$

The second integral, containing the additional cosine term, is easily rewritten as

$$\frac{1}{2}\int_0^a \cos (2\pi x/L)\, \mathrm{d}x - \frac{1}{2}\int_0^a \cos (2m\pi x/a) \cos (2\pi x/L)\, \mathrm{d}x. \tag{4.2.20}$$

Clearly, the integral in the first term evaluates to

$$\frac{L}{2\pi} \sin (2\pi a/L)\,. \tag{4.2.21}$$

Using the cosine angle addition and subtraction identities, the integral in the second term is identical to

$$\frac{1}{2}\int_0^a \left[\cos \left(2\pi x \left(m/a - 1/L\right)\right) + \cos \left(2\pi x \left(m/a + 1/L\right)\right)\right] \mathrm{d}x, \tag{4.2.22}$$

which is straightforwardly evaluated as

$$\frac{\sin\left(2\pi\left(m-\frac{a}{L}\right)\right)}{4\pi\left(\frac{m}{a}-\frac{1}{L}\right)}+\frac{\sin\left(2\pi\left(m+\frac{a}{L}\right)\right)}{4\pi\left(\frac{m}{a}+\frac{1}{L}\right)}. \tag{4.2.23}$$

Note that, because $m$ is an integer,[5]

$$\sin\left[2\pi\left(m\pm\frac{a}{L}\right)\right]=\pm\sin\left(2\pi a/L\right). \tag{4.2.24}$$

It follows from (4.2.20–4.2.24) that the second integral (4.2.20) is

$$\frac{1}{2}\left(\frac{L}{2\pi}\sin\left(2\pi a/L\right)-\frac{a^2}{2\pi L}\left[\frac{\sin\left(2\pi a/L\right)}{(a/L)^2-m^2}\right]\right). \tag{4.2.25}$$

Combining the two integrals, we obtain

$$\omega_{m,1}^{(1)}=-\omega_{m,\lambda}^{(0)}\frac{1}{2a}\left[\frac{a}{2}+\frac{1}{2}\left(\frac{L}{2\pi}\sin\left(2\pi a/L\right)-\frac{1}{2}\frac{a^2}{\pi L}\left[\frac{\sin\left(2\pi a/L\right)}{(a/L)^2-m^2}\right]\right)\right]. \tag{4.2.26}$$

This expression is easily simplified to

$$\omega_{m,1}^{(1)}=-\frac{1}{4}\omega_{m,1}^{L(0)}\left[1+\frac{\sin\left(2\pi a/L\right)}{2\pi}\frac{Lm^2/a}{(a/L)^2-m^2}\right], \tag{4.2.27}$$

giving the first order change in the frequency in the left most portion of the piston for the first polarisation, except when $m=0$, in which case $\omega_{m,1}^{L\,(1)}=0$.

#### 4.2.3.2 Frequency Perturbation of the Second Polarisation

We now turn our attention to the perturbation in the frequency of the second polarisation. For modes $m>0$:

$$\omega_{m,2}^{(1)}=-\frac{\omega_{m,2}^{(0)}}{2a}\frac{1}{k_\parallel^2+k_x^2}\int_0^a\left(k_\parallel^2\cos^2\left(k_x x\right)+k_x^2\sin^2\left(k_x x\right)\right)\left[1-\cos\left(2\pi x/L\right)\right]\mathrm{d}x, \tag{4.2.28}$$

[5] The identity is $\sin\left[2\pi\left(m\pm\frac{a}{L}\right)\right]=\sin\left(2\pi m\right)\cos\left(2\pi a/L\right)\pm\cos\left(2\pi m\right)\sin\left(2\pi a/L\right)$.

recalling that $k_x = m\pi/a$. Again, we can break the integral up into separate contributions. The first can be stated without calculation:

$$\int_0^a \left(k_\parallel^2 \cos^2(m\pi x/a) + k_x^2 \sin^2(m\pi x/a)\right) \mathrm{d}x = \frac{a}{2}\left(k_\parallel^2 + k_x^2\right). \tag{4.2.29}$$

The second integral is composed of two terms:

$$k_\parallel^2 \int_0^a \cos^2(m\pi x/a)\cos(2\pi x/L)\,\mathrm{d}x + k_x^2 \int_0^a \sin^2(m\pi x/a)\cos(2\pi x/L)\,\mathrm{d}x. \tag{4.2.30}$$

The second term has already been computed (4.2.25). The first term can be decomposed into

$$\frac{1}{2}\int_0^a \cos(2\pi x/L)\,\mathrm{d}x + \frac{1}{2}\int_0^a \cos(2m\pi x/a)\cos(2\pi x/L)\,\mathrm{d}x. \tag{4.2.31}$$

Both integrals were computed before (4.2.21, 4.2.22). Equation (4.2.31) is therefore equal to

$$\frac{1}{2}\left(\frac{L}{2\pi}\sin(2\pi a/L) + \frac{a^2}{2\pi L}\left[\frac{\sin(2\pi a/L)}{(a/L)^2 - m^2}\right]\right), \tag{4.2.32}$$

which is the same expression as (4.2.25), but for a change of sign in the second term. It follows after a little simplification that (4.2.30) is

$$\frac{L}{4\pi}\sin(2\pi a/L)\left(k_x^2 + k_\parallel^2\right) + \left(k_\parallel^2 - k_x^2\right)\frac{a^2}{4\pi L}\left[\frac{\sin(2\pi a/L)}{(a/L)^2 - m^2}\right]. \tag{4.2.33}$$

The total integral in (4.2.28) is now determined. From (4.2.29) to (4.2.33) we obtain

$$\omega_{m,2}^{(1)} = -\omega_{m,2}^{(0)} \cdot \frac{1}{2a}\left[\frac{a}{2} - \frac{L}{4\pi}\sin(2\pi a/L) - \frac{k_\parallel^2 - k_x^2}{k_\parallel^2 + k_x^2}\frac{a^2}{4\pi L}\left[\frac{\sin(2\pi a/L)}{(a/L)^2 - m^2}\right]\right]. \tag{4.2.34}$$

The eigenfrequency of the second polarisation thus gets shifted by

$$\omega_{m,2}^{L\,(1)} = -\frac{1}{4}\omega_{m,2}^{L\,(0)}\left\{1 - \frac{\sin(2\pi a/L)}{2\pi}\left[\frac{L}{a}(1-\delta_{m0}) + \left(\frac{k_\parallel^2 - (\frac{m\pi}{a})^2}{k_\parallel^2 + (\frac{m\pi}{a})^2}\right)\frac{a/L}{(a/L)^2 - m^2}\right]\right\}, \tag{4.2.35}$$

where a Kronecker delta function has been added to give the correct behaviour at $m = 0$. After exchanging $a \to L - a$ in (4.2.27) and (4.2.35), we obtain the expressions for the right-hand side of the cavity.

## 4.2.4 Field Energy Perturbations

The rate of change in energy per unit area due to the inhomogeneity of the medium is computed through inserting (4.2.27) and (4.2.35) into (4.2.5). This involves performing integrations of the form

$$I^s_{m,\lambda} = \int_0^\infty k_\| dk_\| \omega^{s\,(1)}_{m,\lambda} e^{-\xi\omega^{s\,(0)}_{m,\lambda}(k_\|)/c}, \tag{4.2.36}$$

where $s \in \{L, R\}$. The $\omega^{s\,(1)}_{m,\lambda}$ are given by (4.2.27) and (4.2.35), with the $s = R$ expressions obtained using the substitution $a \to L - a$.

### 4.2.4.1 Field Energy of the First Polarisation

The eigenfrequency shifts for polarisation $\lambda = 1$ depends on $k_\|$ only through $\omega^{s\,(0)}_{m,1}$, i.e. in the same way as the unperturbed eigenfrequencies. For the purpose of 'bookkeeping', we will write

$$\omega^{L(1)}_{m,1} = \omega^{L(0)}_{m,1}\delta\omega^{L(1)}_{m,1}, \quad \delta\omega^{L(1)}_{m,1} = -\frac{1}{4}\left[1 + \frac{\sin(2\pi a/L)}{2\pi}\frac{Lm^2/a}{(a/L)^2 - m^2}\right]. \tag{4.2.37}$$

It follows that

$$\begin{aligned} I_{m,1} &= \delta\omega^{(1)}_{m,1} \int_0^\infty k_\| \mathrm{d}k_\| \, \omega^{(0)}_{m,1} \mathrm{e}^{-\xi\omega^{(0)}_{m,1}/c} \\ &= c\,\delta\omega^{(1)}_{m,1} \int_0^\infty k_\| \mathrm{d}k_\| \sqrt{k_x^2 + k_\|^2}\, \mathrm{e}^{-\xi\sqrt{k_x^2+k_\|^2}}. \end{aligned} \tag{4.2.38}$$

More concisely:

$$I_{m,1} = -c\,\delta\omega^{(1)}_{m,1} \frac{\partial}{\partial\xi} \int_0^\infty k_\| \mathrm{d}k_\| \, \mathrm{e}^{-\xi\sqrt{k_x^2+k_\|^2}}. \tag{4.2.39}$$

Using the identity

$$k_{\|} \mathrm{e}^{-\xi\sqrt{k_x^2+k_{\|}^2}} = -\frac{\partial}{\partial k_{\|}}\left[\left(\frac{\sqrt{k_x^2+k_{\|}^2}}{\xi}+\frac{1}{\xi^2}\right)\mathrm{e}^{-\xi\sqrt{k_x^2+k_{\|}^2}}\right] \tag{4.2.40}$$

it follows that

$$\int_0^\infty k_{\|}\mathrm{d}k_{\|}\,\mathrm{e}^{-\xi\sqrt{k_x^2+k_{\|}^2}} = -\left(\frac{1}{\xi}\frac{\partial}{\partial\xi}-\frac{1}{\xi^2}\right)\mathrm{e}^{-\xi k_x}, \tag{4.2.41}$$

and we may write

$$I_{m,1} = -\frac{c}{4}\left[1+\frac{\sin(2\pi a/L)}{2\pi}\frac{Lm^2/a}{(a/L)^2-m^2}\right]\frac{\partial}{\partial\xi}\left[\left(\frac{1}{\xi}\frac{\partial}{\partial\xi}-\frac{1}{\xi^2}\right)\mathrm{e}^{-\xi k_x}\right]. \tag{4.2.42}$$

Properly, there is no contribution for $m = 0$, therefore the integral (4.2.36) for this polarisation is

$$\begin{aligned} I_{m,1}^L &= -\frac{c}{4}(1-\delta_{m0})\left[1+\frac{\sin(2\pi a/L)}{2\pi}\frac{Lm^2/a}{(a/L)^2-m^2}\right] \\ &\times\left(\frac{2}{\xi^3}-\frac{2}{\xi^2}\frac{\partial}{\partial\xi}+\frac{1}{\xi}\frac{\partial^2}{\partial\xi^2}\right)e^{-\xi m\pi/a}. \end{aligned} \tag{4.2.43}$$

#### 4.2.4.2 Field Energy of the Second Polarisation

The second polarisation $\lambda = 2$ has a slightly more complicated dependence on $k_{\|}$, but there is nothing fundamentally different about performing the integrals. Again, we begin by setting

$$\omega_{m,2}^{L(1)} = \omega_{m,2}^{L(0)}\delta\omega_{m,2}^{L(1)}(k_{\|}), \tag{4.2.44}$$

with

$$\delta\omega_{m,2}^{L(1)}(k_{\|}) = -\frac{1}{4}\left\{1-\frac{\sin(2\pi a/L)}{2\pi}\left[\frac{L}{a}(1-\delta_{m0})+\frac{k_{\|}^2-\left(\frac{m\pi}{a}\right)^2}{k_{\|}^2+\left(\frac{m\pi}{a}\right)^2}\frac{a/L}{(a/L)^2-m^2}\right]\right\}. \tag{4.2.45}$$

As before (4.2.39), we can write the energy integral in the form

$$I_{m,2} = -c\,\frac{\partial}{\partial\xi}\int\limits_0^\infty k_\parallel \mathrm{d}k_\parallel\;\delta\omega^{(1)}_{m,2}(k_\parallel)\,\mathrm{e}^{-\xi\sqrt{k_x^2+k_\parallel^2}}, \tag{4.2.46}$$

only here we must include the perturbation term within the integral. We will separate (4.2.46) into two integrals and solve them independently:

$$I_{m,2} = I_{m,2a} + I_{m,2b}, \tag{4.2.47}$$

where

$$I_{m,2a} = \frac{c}{4}\left[1 - \frac{L}{a}(1-\delta_{m0})\,\frac{\sin(2\pi a/L)}{2\pi}\right]\frac{\partial}{\partial\xi}\left\{\int\limits_0^\infty k_\parallel \mathrm{d}k_\parallel\,\mathrm{e}^{-\xi\sqrt{k_x^2+k_\parallel^2}}\right\}, \tag{4.2.48}$$

$$I_{m,2b} = -\frac{c}{4}\left[\frac{\sin(2\pi a/L)}{2\pi}\,\frac{a/L}{(a/L)^2-m^2}\right]\int\limits_0^\infty k_\parallel \mathrm{d}k_\parallel\left\{\frac{k_\parallel^2-\left(\frac{m\pi}{a}\right)^2}{\sqrt{k_\parallel^2+\left(\frac{m\pi}{a}\right)^2}}\right\}\mathrm{e}^{-\xi\sqrt{k_x^2+k_\parallel^2}}. \tag{4.2.49}$$

The integral in (4.2.48) has already been solved. From (4.2.41), it follows that

$$I_{m,2a} = -\frac{c}{4}\left[1 - \frac{L}{a}(1-\delta_{m0})\,\frac{\sin(2\pi a/L)}{2\pi}\right]\left[\left(\frac{2}{\xi^3} - \frac{2}{\xi^2}\frac{\partial}{\partial\xi} + \frac{1}{\xi}\frac{\partial^2}{\partial\xi^2}\right)\mathrm{e}^{-\xi k_x}\right]. \tag{4.2.50}$$

In solving the integral in (4.2.49), we note that, in analogy with (4.2.40), the integrand can be written in the form

$$-\frac{\partial}{\partial k_\parallel}\left[\left(\frac{2}{\xi^3} + \frac{2\sqrt{k_x^2+k_\parallel^2}}{\xi^2} - \frac{\left(k_x^2-k_\parallel^2\right)}{\xi}\right)\mathrm{e}^{-\xi\sqrt{k_x^2+k_\parallel^2}}\right]. \tag{4.2.51}$$

Thus we determine that

$$I_{m,2b} = -\frac{c}{4}\left[\frac{\sin(2\pi a/L)}{2\pi}\,\frac{a/L}{(a/L)^2-m^2}\right]\left(\frac{2}{\xi^3} - \frac{2}{\xi^2}\frac{\partial}{\partial\xi} - \frac{1}{\xi}\frac{\partial^2}{\partial\xi^2}\right)\mathrm{e}^{-\xi k_x}. \tag{4.2.52}$$

Combining the two terms $I_{m,2} = I_{m,2a} + I_{m,2b}$, we obtain the total perturbation in the contribution of the second polarisation to the ground-state energy:

$$I^L_{m,2} = -\frac{c}{4}\Bigg\{\left[1 - \frac{\sin(2\pi a/L)}{2\pi}\left(\frac{L}{a}(1-\delta_{m0}) + \frac{a/L}{(a/L)^2 - m^2}\right)\right] \\ \times\left(\frac{2}{\xi^3} - \frac{2}{\xi^2}\frac{\partial}{\partial\xi} + \frac{1}{\xi}\frac{\partial^2}{\partial\xi^2}\right) + \frac{\sin(2\pi a/L)}{2\pi} \\ \times\left(\frac{2a/L}{(a/L)^2 - m^2}\right)\frac{1}{\xi}\frac{\partial^2}{\partial\xi^2}\Bigg\} e^{-\xi m\pi/a}. \tag{4.2.53}$$

This can be rewritten in the form

$$I^L_{m,2} = -\frac{c}{4}\Bigg\{\left[1 - \frac{\sin(2\pi a/L)}{2\pi}\left(\frac{L}{a} + \frac{a/L}{(a/L)^2 - m^2}\right)\right] \\ \times\left(\frac{2}{\xi^3} - \frac{2}{\xi^2}\frac{\partial}{\partial\xi} + \frac{1}{\xi}\frac{\partial^2}{\partial\xi^2}\right) e^{-\xi m\pi/a} \\ + \left[\frac{\sin(2\pi a/L)}{2\pi}\left(\frac{2a/L}{(a/L)^2 - m^2}\right)\frac{1}{\xi}\frac{\partial^2}{\partial\xi^2}\right] e^{-\xi m\pi/a} \\ + \delta_{m0}\frac{\sin(2\pi a/L)}{2\pi a/L}\frac{2}{\xi^3}\Bigg\}. \tag{4.2.54}$$

After substituting $a \to L - a$, one obtains $I^R_{m,1}$ and $I^R_{m,2}$.

#### 4.2.4.3 Total Field Energy

The total change in the energy includes the contributions of both polarisations, on both sides of the cavity:

$$\frac{1}{A}\frac{\partial E}{\partial\alpha} = \frac{\hbar}{4\pi}\frac{\partial}{\partial\xi}\sum_m \xi\left[I_{m,1} + I_{m,2}\right] + \{a \to L - a\}. \tag{4.2.55}$$

On adding the two contributions we obtain:

$$I_{m,1} + I_{m,2} = -\frac{c}{4}\Bigg\{\left[1 + \frac{\sin(2\pi a/L)}{2\pi}\frac{Lm^2/a}{(a/L)^2 - m^2}\right]\left(\frac{2}{\xi^3} - \frac{2}{\xi^2}\frac{\partial}{\partial\xi} + \frac{1}{\xi}\frac{\partial^2}{\partial\xi^2}\right)\mathrm{e}^{-\xi k_x} \\ + \left[1 - \frac{\sin(2\pi a/L)}{2\pi}\left(\frac{L}{a} + \frac{a/L}{(a/L)^2 - m^2}\right)\right]$$

$$\times\left(\frac{2}{\xi^3}-\frac{2}{\xi^2}\frac{\partial}{\partial\xi}+\frac{1}{\xi}\frac{\partial^2}{\partial\xi^2}\right)\mathrm{e}^{-\xi k_x}+\frac{\sin(2\pi a/L)}{2\pi}$$
$$\times\left(\frac{2a/L}{(a/L)^2-m^2}\right)\frac{1}{\xi}\frac{\partial^2}{\partial\xi^2}\mathrm{e}^{-\xi k_x}+\delta_{m0}\left(\frac{\sin(2\pi a/L)}{2\pi a/L}-1\right)\frac{2}{\xi^3}\bigg\}. \tag{4.2.56}$$

This expression can be simplified by factorising, and by noting that

$$\left\{\frac{Lm^2/a}{(a/L)^2-m^2}-\frac{L}{a}-\frac{a/L}{(a/L)^2-m^2}\right\}=-\frac{2L}{a}, \tag{4.2.57}$$

leaving

$$I_{m,1}+I_{m,2}=-\frac{c}{4}\left\{2\left[1-\frac{\sin(2\pi a/L)}{2\pi a/L}\right]\left(\frac{2}{\xi^3}-\frac{2}{\xi^2}\frac{\partial}{\partial\xi}+\frac{1}{\xi}\frac{\partial^2}{\partial\xi^2}\right)\mathrm{e}^{-\xi k_x}\right.$$
$$\left.+\frac{\sin(2\pi a/L)}{2\pi}\left(\frac{2a/L}{(a/L)^2-m^2}\right)\frac{1}{\xi}\frac{\partial^2}{\partial\xi^2}\mathrm{e}^{-\xi k_x}-\epsilon\delta_{m0}\right\}, \tag{4.2.58}$$

where

$$\epsilon=\left(1-\frac{\sin(2\pi a/L)}{2\pi a/L}\right)\frac{2}{\xi^3}. \tag{4.2.59}$$

Thus the rate of change of the ground-state energy, as the amplitude of $\delta\epsilon$ is increased from zero to $\delta\alpha$, is given by

$$\frac{1}{A}\frac{\partial E}{\partial\alpha}=-\frac{\hbar c}{16\pi}\frac{\partial}{\partial\xi}\sum_m\xi\left\{\left[2\left[1-\frac{\sin(2\pi a/L)}{2\pi a/L}\right]\left(\frac{2}{\xi^3}-\frac{2}{\xi^2}\frac{\partial}{\partial\xi}+\frac{1}{\xi}\frac{\partial^2}{\partial\xi^2}\right)\mathrm{e}^{-\xi k_x}\right]\right.$$
$$\left.+\frac{\sin(2\pi a/L)}{2\pi}\left(\frac{2a/L}{(a/L)^2-m^2}\right)\frac{1}{\xi}\frac{\partial^2}{\partial\xi^2}\mathrm{e}^{-\xi k_x}-\epsilon\delta_{m0}\right\}+\{a\to L-a\}. \tag{4.2.60}$$

### 4.2.5 Removing the Regularisation

Until the regularisation is removed, the energy is cut off arbitrarily and expression (4.2.60) is not physically informative. Previously, our procedure has been to extract the free energy of the system—that is, to isolate the mechanical part of the ground-state energy of the system that depends on the distance between the plates and the moving piston, taking the parameter $\xi\to 0$ (i.e. removing the cut-off). It is necessary, then, to formulate (4.2.60) in a way that renders conspicuous its behaviour in this limit. First we note that

$$e^{\xi\pi/2a}\text{cosech}\,(\xi\pi/2a) = \frac{2e^{\xi\pi/2a}}{e^{\xi\pi/2a} - e^{-\xi\pi/2a}} = \frac{2}{1 - e^{-\xi\pi/a}} = 2\sum_m e^{-\xi m\pi/a}. \tag{4.2.61}$$

Also, for reasons which will become apparent, it is convenient to use the partial fraction decomposition

$$\frac{1}{m^2 - (a/L)^2} = -\frac{L}{2a}\left(\frac{1}{m + (a/L)} - \frac{1}{m - (a/L)}\right). \tag{4.2.62}$$

We can then write (4.2.60) in the form

$$\begin{aligned}\frac{1}{A}\frac{\partial E}{\partial \alpha} = -\frac{\hbar c}{16\pi}\frac{\partial}{\partial \xi}\xi\Bigg\{&\left(1 - \frac{\sin(2\pi a/L)}{2\pi a/L}\right)\left(\frac{2}{\xi^3} - \frac{2}{\xi^2}\frac{\partial}{\partial \xi} + \frac{1}{\xi}\frac{\partial^2}{\partial \xi^2}\right)e^{\xi\pi/2a}\text{cosech}\,(\xi\pi/2a)\\ &+\frac{\sin(2\pi a/L)}{2\pi}\frac{1}{\xi}\frac{\partial^2}{\partial \xi^2}\sum_{m=0}\left[\frac{e^{-\xi m\pi/a}}{m + (a/L)} - \frac{e^{-\xi m\pi/a}}{m - (a/L)}\right] - \epsilon\Bigg\}\\ &+ \{a \to L - a\}.\end{aligned} \tag{4.2.63}$$

We note that, in addition to terms proportional to a geometric series, the resulting sum over $m$ contains terms of the form,

$$\Phi(e^{-\xi\pi/a}, 1, v) = \sum_{m=0}^{\infty} \frac{e^{-\xi m\pi/a}}{m + v}. \tag{4.2.64}$$

The quantity $\Phi$ is known as the Lerch function, and its properties have been investigated elsewhere [6, 7]. Applying the notation of (4.2.64), we can perform the second summation, with the result that

$$\begin{aligned}\frac{1}{A}\frac{\delta \tilde{E}}{\delta \alpha} = -\frac{\hbar c}{16\pi}\frac{\partial}{\partial \xi}\xi\Bigg[&\left(1 - \frac{\sin(2\pi a/L)}{2\pi a/L}\right)\left(\frac{2}{\xi^3} - \frac{2}{\xi^2}\frac{\partial}{\partial \xi} + \frac{1}{\xi}\frac{\partial^2}{\partial \xi^2}\right)e^{\xi\pi/2a}\text{cosech}(\xi\pi/2a)\\ &+\frac{\sin(2\pi a/L)}{2\pi}\frac{1}{\xi}\frac{\partial^2}{\partial \xi^2}\left(\Phi(e^{-\xi\pi/a}, 1, a/L) - \Phi(e^{-\xi\pi/a}, 1, -a/L)\right) - \epsilon\Bigg]\\ &+ \{a \to L - a\}.\end{aligned} \tag{4.2.65}$$

Again we return to considering the limit $\xi \to 0$, making use of the series expansion of the Lerch function [7]:

$$\Phi(z, n, v)\, z^v = \sum_{\substack{m=0\\ m\neq n-1}}^{\infty}\left\{\zeta(n - m, v)\frac{\log^m(z)}{m!} + \left[\psi(n) - \psi(v) - \log(-\log(z))\right]\frac{\log^{n-1}(z)}{(n-1)!}\right\},$$

where $\psi$ is the digamma function,[6] and $\zeta(z, q)$ is the Hurwitz zeta function.[7] Consider the $\Phi$ terms in $\delta E_1$. For $z = \mathrm{e}^{-\lambda\pi/a}$ and $n = 1$:

$$\log^m(z) = \log^m(\mathrm{e}^{-\lambda\pi/a}) = (-1)^m\left(\frac{\lambda\pi}{a}\right)^m, \quad \frac{\log^{n-1}(z)}{(n-1)!} = \frac{1}{0!} = 1, \tag{4.2.66}$$

$$-\log(-\log(z)) = -\log\left(\frac{\lambda\pi}{a}\right) = -\log(\lambda) - \log(\pi/a), \tag{4.2.67}$$

Also, $\psi(1)$ is equal to $-\gamma$, where $\gamma$ is Euler's constant. Inserting (4.2.66) and (4.2.67) into (4.2.5), we obtain the expression

$$\Phi(\mathrm{e}^{-\xi\pi/a}, 1, v) = \mathrm{e}^{\xi\pi v/a}\left\{\sum_{k=1}^{\infty}(-1)^k\zeta(1-k, v)\frac{(\xi\pi/a)^k}{k!} + \gamma - \psi(v) - \ln(\xi\pi/a)\right\}. \tag{4.2.68}$$

The Hurwitz zeta function is related to the Bernoulli polynomials when the first argument is a negative integer [6]:

$$\zeta(-n, v) = -\frac{B_{n+1}(v)}{n+1}, \tag{4.2.69}$$

where the first three Bernoulli polynomials are

$$B_1(v) = v - \frac{1}{2},$$
$$B_2(v) = v^2 - v + \frac{1}{6},$$
$$B_3(v) = v^3 - \frac{3}{2}v^2 + \frac{1}{2}v.$$

[6] The digamma function is the logarithmic derivative of the gamma function:

$$\psi(x) = \frac{\mathrm{d}}{\mathrm{d}x}\ln\Gamma(x), \quad \Gamma(x) = \int_0^\infty y^{x-1}e^{-y}\mathrm{d}y.$$

[7] The Hurwitz zeta function is defined for complex arguments $z$ and $q$, with $\mathrm{Re}(z) > 1$ and $\mathrm{Re}(q) > 1$,

$$\zeta(z, q) = \sum_{n=0}^{\infty}\frac{1}{(q+n)^z}.$$

The Lerch function can now be written[8]

$$\Phi(\mathrm{e}^{-\xi\pi/a}, 1, v) = -\mathrm{e}^{\xi\pi v/a}\left(\gamma + \psi(v) + \ln(\xi\pi/a) + \sum_{m=0}^{\infty}(-1)^{m+1}\frac{B_{m+1}(v)}{m+1}\frac{(\xi\pi/a)^{m+1}}{(m+1)!}\right). \tag{4.2.70}$$

In the regime of interest—where $\xi \ll 1$ and $0 < v < 1$—the expansion (4.2.70) is well behaved. From (4.2.70) we find, after neglecting positive powers[9] of $\xi$,

$$\begin{aligned}\frac{1}{\xi}\frac{\partial^2}{\partial\xi^2}\left[\Phi(e^{-\xi\pi/a}, 1, v) - \Phi(e^{-\xi\pi/a}, 1, -v)\right] &\sim -\frac{2\pi v}{a\xi^2} \\ &- 2\left(\frac{\pi v}{a}\right)^3 \log(\xi\pi/a) - \frac{\pi^2 v}{a^2\xi}\left[1 + v(\psi(v) - \psi(-v))\right] \\ &- \left(\frac{\pi}{a}\right)^3 v\left\{v^2[2(\gamma-1) + \psi(v) + \psi(-v)] + 1/6\right\}.\end{aligned} \tag{4.2.71}$$

Note also that

$$e^{\xi\pi/2a}\mathrm{cosech}\,(\xi\pi/2a) \approx \frac{2a}{\xi\pi} + 1 + \frac{1}{3}\left(\frac{\xi\pi}{2a}\right) - \frac{1}{45}\left(\frac{\xi\pi}{2a}\right)^3. \tag{4.2.72}$$

Then

$$\left(\frac{2}{\xi^3} - \frac{2}{\xi^2}\frac{\partial}{\partial\xi} + \frac{1}{\xi}\frac{\partial^2}{\partial\xi^2}\right)e^{\xi\pi/2a}\mathrm{cosech}\,(\xi\pi/2a) = -\frac{\pi^3}{180a^3} + \frac{12a}{\pi\xi^4} + \frac{2}{\xi^3}. \tag{4.2.73}$$

Applying (4.2.71) and (4.2.72) to (4.2.65) yields the following expression for the rate of change of energy with respect to $\alpha$:

$$\begin{aligned}\frac{1}{A}\frac{\delta\tilde{E}}{\delta\alpha} \approx \frac{\hbar c}{4\pi}\Bigg\{&\left(1 - \frac{\sin(2\pi a/L)}{2\pi a/L}\right)\left(\frac{9a}{\xi^4\pi} + \frac{1}{\xi^3} + \frac{\pi^3}{720a^3}\right) \\ &- \frac{\sin(2\pi a/L)}{4\pi}\Bigg[\frac{\pi}{L\xi^2} - \frac{\pi^3}{L^3}\left[\log(\xi\pi/a) + 1\right] \\ &\qquad - \frac{\pi^3}{2L^3}(2(\gamma-1) + \psi(a/L) + \psi(-a/L)) - \frac{\pi^3}{12a^2L}\Bigg] \\ &- \left(1 - \frac{\sin(2\pi a/L)}{2\pi a/L}\right)\frac{1}{\xi^3}\Bigg\} + \{a \to L - a\}.\end{aligned} \tag{4.2.74}$$

---

[8] An examination of [7] may suggest that the expansion (4.2.70) does not apply to the case when the middle index equals unity. However [6] confirms that it does in fact hold in this case; the expansion can be simplified to a hypergeometric function, but we do not require this simplification. In addition, we have verified (4.2.70) by direct numerical evaluation, comparing it to the results of (4.2.64).

[9] The reader should recall that we are anticipating the limit $\xi \to 0$.

The terms in $1/\xi^3$ are clearly seen to cancel, but the energy expression for the left side of the cavity contains a contribution which diverges in the limit $\xi \to 0$ given by

$$\left(1 - \frac{\sin(2\pi a/L)}{2\pi a/L}\right)\frac{9a}{\xi^4\pi} - \frac{\sin(2\pi a/L)}{4\pi}\left(\frac{\pi}{L\xi^2} - \frac{\pi^3}{L^3}\log(\xi)\right). \quad (4.2.75)$$

However, a similarly diverging contribution to the energy from the right side of the cavity is obtained by applying the transformation $a \to L-a$ to the expression above, and the remaining divergences are seen to cancel apart from one term $9L/\xi^4\pi$. The total modification to the Casimir energy (4.2.74) containing the contributions from both sides of the cavity can consequently be rewritten in the form

$$\frac{1}{A}\frac{\delta\tilde{E}}{\delta\alpha} \approx \frac{\hbar c}{4\pi}\left(\frac{9L}{\xi^4\pi} + \delta\right), \quad (4.2.76)$$

where $\delta$ consists of the remaining finite terms in the Casimir energy:

$$\begin{aligned}\delta = &-\frac{1}{4\pi}\sin(2\pi a/L)\left(-\frac{\pi^3}{2L^3}\left[\psi(a/L) + \psi(-a/L) - \psi(1-a/L) - \psi(a/L-1)\right]\right)\\ &-\frac{1}{4\pi}\sin(2\pi a/L)\left(\frac{\pi^3}{L^3}\log(a) - \frac{\pi^3}{12L}\frac{1}{a^2}\right)\\ &+\frac{1}{4\pi}\sin(2\pi a/L)\left(\frac{\pi^3}{L^3}\log(L-a) - \frac{\pi^3}{12L}\frac{1}{(L-a)^2}\right)\\ &+\left(1 - \frac{1}{2\pi}\frac{\sin(2\pi a/L)}{a/L}\right)\frac{\pi^3}{720a^3} + \left(1 + \frac{1}{2\pi}\frac{\sin(2\pi a/L)}{1-a/L}\right)\frac{\pi^3}{720(L-a)^3}. \quad (4.2.77)\end{aligned}$$

This expression for the finite contribution to the Casimir energy can be further simplified by exploiting the special properties of the digamma function. Using the definition of the gamma function and integration by parts, it is trivial to show that

$$\Gamma(x+1) = x\Gamma(x), \quad \Gamma(x-1) = \frac{\Gamma(x)}{x-1}. \quad (4.2.78)$$

It follows straightforwardly from the definition of the digamma function that

$$\psi(1-x) = \psi(-x) - \frac{1}{x}, \quad \psi(x-1) = \psi(x) - \frac{1}{x-1}, \quad (4.2.79)$$

thus establishing the identity

$$\psi(x) + \psi(-x) - \psi(1-x) - \psi(x-1) \equiv \frac{1}{x} + \frac{1}{x-1}. \quad (4.2.80)$$

Applying Eqs. (4.2.80) to (4.2.77) and simplifying, we obtain

$$\delta = \frac{\pi^2}{4L^3}\sin(2\pi a/L)\left[\frac{1}{2}\left(\frac{L}{a}\right) + \frac{1}{12}\left(\frac{L}{a}\right)^2 - \frac{1}{360}\left(\frac{L}{a}\right)^4 - \log(a)\right]$$
$$+ \frac{\pi^3}{720a^3} + \{a \to L - a\}. \tag{4.2.81}$$

The term $9L/\xi^4\pi$ in (4.2.76) is divergent in the limit $\xi \to 0$, but vanishes in the force-derivative. It follows that the part of the Casimir energy associated with the moveable mirror is cutoff-independent: the force on the mirror can be stated exactly when computed from a summation over modes.

### 4.2.6 Special Cases

Two special cases are worth noting. First, we can effectively collapse the piston geometry into a simple cavity by taking $L \to \infty$. All but two terms clearly vanish in the limit. Noting that

$$\lim_{L\to\infty} L\sin(2\pi a/L) = 2\pi a, \tag{4.2.82}$$

we find that the two remaining terms in (4.2.81) cancel and $\delta = 0$. There is no modification to the Casimir force. This is to be expected: in this limit, we have effectively removed the perturbation to the profile $f(x) \to 0$ and reduced the material between the mirrors to vacuum.

Secondly, it is also worth noting that when the mirror is positioned exactly at the centre of the cavity ($a = L/2$), where the gradient of the peturbation profile is zero, the rate of change of the energy reduces to the much simpler expression

$$\frac{1}{A}\frac{\delta\tilde{E}}{\delta\alpha} \approx \frac{\hbar c}{4\pi}\left(\frac{9L}{\xi^4\pi} + \frac{\pi^3}{720a^3} + \frac{\pi^3}{720(L-a)^3}\right). \tag{4.2.83}$$

## 4.3 Summary Remarks

### 4.3.1 The Cut-off Independence of the Casimir Energy

The investigation in this chapter demonstrates that the ground-state energy of an inhomogeneous system can be determined from a simple energy mode summation, using standard techniques of regularisation: we found an exact analytic expression for the Casimir energy of a piston filled with an inhomogeneous medium, in which the single diverging part of the energy remains independent of the positions of the mirrors and does not modify the value of the force. There does not seem to be any reason to suppose that this pleasing result is a peculiar feature of this profile; it should be possible to generalise this result for arbitrary profiles (for example, by rewriting the profile as a Fourier series decomposition, and proceeding along the same lines as above). This suggests that it should be possible to determine Casimir quantities in inhomogeneous systems without reference to their microphysical properties, despite some suggestions to the contrary [3, 8, 9]. Our result is consistent with a numerical regularisation scheme recently proposed in [10], in which the Casimir force in an inhomogeneous medium is determined by forming a Laurent expansion of the energy (or pressure) in powers of the regularising parameter[10] (in our case, in terms of $\xi$).

### 4.3.2 Possible Objections

It must be acknowledged, however, that in this investigation we have employed a rather basic model of a dielectric medium: like Casimir [11], we consider only a simple energy summation of the field modes between perfect mirrors. We make the artificial assumption that $\epsilon$ and $\mu$ are independent of frequency; the propagation speed of light is simply varied from point to point, and not as a function of frequency. Under these conditions, we have found that the Casimir force remains well-behaved in the limit where the regularisation (cut-off) tends to infinite frequency, and we can obtain an exact expression for the zero point force that is independent of the choice of cut-off in the energy summation. Nevertheless, real media are both dispersive and dissipative, and to properly account for these phenomena, and contend with any doubts that may arise due to the simplicity of our model, we require Lifshitz theory [12–15]. It is to this more sophisticated apparatus that we must now turn.

## References

1. M. Bordag, G.L. Klimchitskaya, U. Mohideen, V.M. Mostepanenko, *Advances in the Casimir Effect* (Oxford University Press, Oxford, 2009)
2. D.A.R. Dalvit, P.W. Milonni, D. Roberts, F. Rosa, *Casimir Physics: Lecture Notes in Physics*, vol. 834 (Springer, Berlin, 2011) (chapter Introduction)
3. S.A.R. Horsley, W.M.R. Simpson, Phys. Rev. A **88**, 013833 (2013)
4. D. Pozar, *Microwave Engineering* (Wiley, New York, 1998)
5. S.G. Johnson, M. Ibanescu, M.A. Skorobogatiy, O. Weisberg, J.D. Joannopoulos, Y. Fink, Phys. Rev. E **65**, 066611 (2002)
6. H. Bateman, *Higher Transcendental Functions*, vol. 1 (McGraw-Hill, New York, 1953)
7. I.S. Gradshteyn, I.M. Ryzhik, *Table of Integrals, Series, and Products* (Elsevier, Amsterdam, 2007)

[10] The regularised quantity is then defined to be the term $c_0$ in the Laurent expansion that is independent of the regularising parameter.

8. U. Leonhardt, W.M.R. Simpson, Phys. Rev. D **84**, 081701 (2011)
9. T.G. Philbin, C. Xiong, U. Leonhardt, Ann. Phys. **325**, 579 (2009)
10. S. Goto, A. Hale, R. Tucker, T. Walton, Phys. Rev. A **85**, 034103 (2012)
11. H.B.G. Casimir, Koninkl. Ned. Akad. Wetenschap. **51**, 793 (1948)
12. E.M. Lifshitz, Zh. Eksp. Teor. Fiz. **29**, 94 (1955)
13. I.E. Dzyaloshinskii, E.M. Lifshitz, L.P. Pitaevskii, Adv. Phys. **10**, 165 (1961)
14. T.G. Philbin, New. J. Phys. **13**, 063026 (2011)
15. E.M. Lifshitz, L.P. Pitaevskii, (Butterworth-Heinemann, Oxford, 2003)

# Chapter 5
# The Divergence of the Casimir Stress

*I shall be telling this with a sigh... Two roads diverged in a wood, and I—I took the one less traveled by*

Robert Frost, *The Road Not Taken*

## 5.1 The Casimir Force in Real Media

In the last chapter, we revisited Casimir's original problem of a cavity formed by two perfect mirrors, making a simple modification: an inhomogeneous medium with a continuously varying permittivity profile was introduced into the cavity. Under these conditions, we found that the Casimir-energy of the system, construed as a simple mode summation, can be stated exactly. If this is the case, we might reasonably expect that a generally finite and physically meaningful result can be obtained for systems embodying small-scale inhomogeneities, without incorporating additional information about the microphysical details of their structure.

Nevertheless, Casimir's model—involving a cavity formed by two perfectly reflecting mirrors, which we adopted in our investigation—represents a highly idealised case. There are no perfect mirrors in nature, and real media are both dispersive and dissipative. To predict the behaviour of a realistic physical system, we require a calculus with the capacity to incorporate these phenomena within its description. As discussed in Chap. 2, Lifshitz theory offers such an apparatus [1] that fits fairly well with experimental results [2, 3].

Our aim in this chapter is to repeat our previous thought-experiment, but using a more realistic model involving dispersive dielectric rather than perfect mirrors.[1] The case of macroscopic materials embedded within a fluid is of experimental

[1] The main results in this chapter were published in [4].



W.M.R. Simpson, *Surprises in Theoretical Casimir Physics*, Springer Theses,
DOI 10.1007/978-3-319-09315-4_5

interest [2], and in this case the Casimir force must be computed using the stress tensor within the fluid. In this problem, we seek the value of the force when the fluid is inhomogeneous—for example, when there is a variation in the density of the fluid.[2]

## 5.2 The Regularised Stress in the Continuum Limit

In Chap. 2, we derived the general form of the stress tensor from which the Casimir forces in a system can be determined. The formalism is written in terms of the electromagnetic Green function, which describes the field produced by charges and currents within the system (2.2.9), (2.2.10). The ground state of the coupled system of electromagnetic field and dielectric is one with non–zero current density within the media [5, 6], consistent with the fluctuation-dissipation theorem [7]. Casimir–Lifshitz forces arise from the interaction of these ground-state currents.

The stress tensor, however, contains the same divergent contribution that appeared in Casimir's original work, and must also be regularised. As we discussed in Sect. 2.3.3, this is typically achieved by subtracting from the total Green function an auxiliary Green function associated with an infinite homogeneous medium [6, 8–11]. One can then compute a finite stress tensor for the system that depends on the dielectric functions of the material at imaginary frequencies. Only then can the force be derived. Both Casimir's and Lifshitz' regularisations give identical results in the limiting case of a cavity sandwiched between perfectly reflecting mirrors.[3]

### *5.2.1 The Stress Tensor for a Rectangular Cavity*

The usual expression for the stress tensor, when applied to a medium that is defined piece–wise along a single axis, is known to be finite.[4] To be explicit, for a region of width $a$ where $\epsilon$ and $\mu$ are homogeneous, the value of the regularised stress tensor at a point $x$ can be written in terms of the reflection coefficients (as opposed to the Green functions) associated with sending $q$–polarised ($q = s, p$) plane waves to the right ($r_{qR}$) and to the left ($r_{qL}$) of this point [10, 12, 13],

$$\sigma_{xx}(x) = 2\hbar c \sum_{q=s,p} \int_0^{\infty} \frac{d\kappa}{2\pi} \int_{\mathbb{R}^2} \frac{d^2\boldsymbol{k}_{\parallel}}{(2\pi)^2} w \frac{r_{qL} r_{qR} e^{-2aw}}{1 - r_{qL} r_{qR} e^{-2aw}}, \tag{5.2.1}$$

[2] For example, sugar dissolved in water under gravity produces an inhomogeneous fluid. This can easily be verified with a laser—light rays entering the fluid become curved.

[3] See Sect. 2.4.2.

[4] Significantly, this is *not* the case for all of the components of the stress tensor: near the boundaries between distinct homogeneous regions, the lateral components of the stress diverge, but the normal component remains finite.

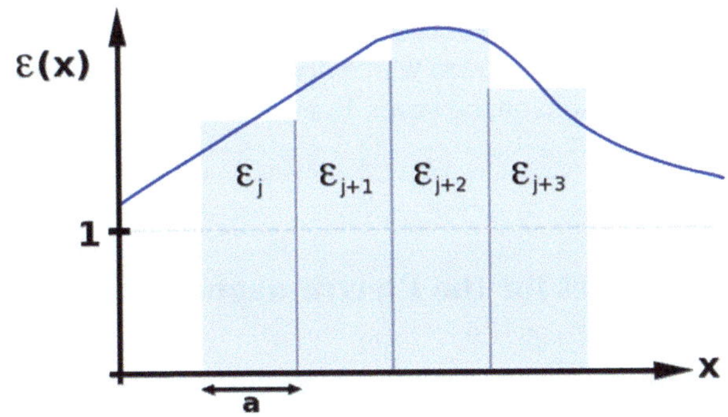


**Fig. 5.1** We consider a medium that is inhomogeneous along $x$, dividing it into $N$ homogeneous slices of width $a$. The local value of the regularised stress tensor (5.2.1) is then investigated within the medium in the limit as $a \rightarrow 0$. For the purposes of illustration only the permittivity $\epsilon(x)$ is shown here. Our analysis holds for both inhomogeneous permittivity and permeability

where $w = (n^2\kappa^2 + k_{\parallel}^2)^{1/2}$, $k_{\parallel} = |\boldsymbol{k}_{\parallel}|$, and $n$ is the value of the refractive index in the homogeneous region surrounding $x$. The reflection coefficients are functions of the imaginary frequency, $\omega = ic\kappa$, the (real) in–plane wave–vector $\boldsymbol{k}_{\parallel}$, and the material parameters of the media to the right and to the left of the homogeneous region. This equation is identical to Eq. (2.4.26) which was derived in Sect. 2.4. The advantage of writing the stress tensor in this form is that the regularisation procedure of Lifshitz theory is automatically implemented [10]. The contributions to the stress arise entirely from inhomogeneities in the system. In this chapter we will investigate the behaviour of (5.2.1) in the limit as the piece–wise definition of the medium (see Fig. 5.1) becomes a continuous function ($a \rightarrow 0$).

### 5.2.2 An Anticipated Divergence

Before we proceed with a more lengthy argument, let us consider (5.2.1) in a cavity of width $a$. This quantity is finite when we integrate over $k_{\parallel}$ due to the exponential decay of the field across the cavity at imaginary frequencies, the rate of decay becoming increasingly rapid as $k_{\parallel}$ increases. Indeed, once $k_{\parallel}$ becomes sufficiently large then the field cannot reach the boundaries of the cavity at all and the reflection coefficients correspondingly tend to zero. However, upon shrinking $a$, this convergence becomes slower, a higher value of $k_{\parallel}$ being required before the field fails to make a round trip across the cavity. Given that a continuous medium can be understood as the limit where $a \rightarrow 0$, and the refractive index contrast between the cavity and the walls becomes infinitesimal, we should ask whether the reflection coefficients vanish fast enough as $a \rightarrow 0$ in order for the stress (5.2.1) to be finite. It seems that they do not: changing variables in (5.2.1) to $\zeta = aw$, and $\xi = ak_{\parallel}$, we find the whole integral multiplied by $a^{-3}$. Meanwhile in this limit the reflection coefficients would in general have contributions linear in $a$, which would still leave a term proportional to $a^{-1}$ within the stress tensor: a term which diverges in the continuum limit, where $a \rightarrow 0$. This remains a suspicion, however. In what follows, we will try to make this argument more precise.

## 5.3 Transfer Matrices for the Electromagnetic Field

To describe the Casimir stress in an inhomogeneous medium, we will slice it into $N$ homogeneous portions of equal width $a$, with $Na = L$. In order to evaluate this quantity, we must be able to determine the 'left' and 'right' reflection coefficients of both polarisations, $r_{L\lambda}$ and $r_{R\lambda}$ ($\lambda = 1, 2$), for any slice of the multilayer structure.[5] To achieve this, we must first develop a formalism for tracking the behaviour of the electromagnetic field throughout this structure. In this case, we will deploy the transfer matrix technique for our analysis of the field [13–16]. In what follows, we will derive transfer matrices suitable to our inquiry.

### 5.3.1 *Single-Interface Transfer Matrix*

We begin by determining the boundary conditions on the electromagnetic field at a sharp interface between two homogeneous half-spaces.

#### 5.3.1.1 Boundary Conditions for the Electric Field

For the electric field, recalling that $\nabla \cdot \mathbf{D} = 0$, we integrate over a volume such that the boundary sits between its upper and lower surface, where

$$\int_V \nabla \cdot \mathbf{D} \, \mathrm{d}V = 0. \tag{5.3.1}$$

We can shrink the walls of the volume so that all the flux of the field enters or leaves through the top and bottom surfaces, and

$$\int_V \nabla \cdot \mathbf{D} \, \mathrm{d}V = \int \mathbf{D} \cdot \mathrm{d}S \implies \mathbf{D_1} \cdot \hat{\mathbf{x}} \Delta S + \mathbf{D}_2 \cdot \left(-\hat{\mathbf{x}} \Delta S\right) = 0. \tag{5.3.2}$$

[5] This technique may be similarly employed to recover the Green function. See Appendix C.

This determines continuity conditions at the boundary for the normal component of **D**:

$$\mathbf{D}_1 \cdot \hat{\mathbf{x}} = \mathbf{D}_2 \cdot \hat{\mathbf{x}}, \tag{5.3.3}$$

where $\hat{\mathbf{x}}$ is the vector normal to the interface. We can derive conditions for the tangential component of the electric field by applying Faraday's law to a small rectangular loop positioned across the boundary:

$$\int_S \nabla \times \mathbf{E}\,\mathrm{d}S = -\frac{\partial}{\partial t}\int \mathbf{B}\,\mathrm{d}S. \tag{5.3.4}$$

Again, we consider the limiting case where the sides are permitted to contract to zero length, preventing any magnetic flux from cutting the surface, so that

$$\int_S \nabla \times \mathbf{E}\,\mathrm{d}S = \oint \mathbf{E}\,\mathrm{d}l = 0 \implies \int_a^b \mathbf{E}_1 \cdot \mathbf{dl} + \int_c^d \mathbf{E}_2 \cdot \mathbf{dl} = 0$$

$$\implies \mathbf{E}_1 \cdot \Delta\mathbf{l} + \mathbf{E}_2 \cdot (-\Delta\mathbf{l}) = 0. \tag{5.3.5}$$

This determines continuity conditions at the boundary for the tangential component of **E**:

$$\mathbf{E}_1 \cdot \hat{\mathbf{t}} = \mathbf{E}_2 \cdot \hat{\mathbf{t}}, \tag{5.3.6}$$

where $\hat{\mathbf{t}}$ is the vector parallel to the interface.

#### 5.3.1.2 Boundary Conditions for the Magnetic Field

We proceed similarly for the magnetic field. Recalling that $\nabla \cdot \mathbf{B} = 0$, we find that

$$\mathbf{B}_1 \cdot \hat{\mathbf{x}} = \mathbf{B}_2 \cdot \hat{\mathbf{x}}. \tag{5.3.7}$$

Recalling that $\nabla \times \mathbf{H} = \frac{\partial \mathbf{D}}{\partial t}$, we also determine that

$$\mathbf{H}_1 \cdot \hat{\mathbf{t}} = \mathbf{H}_2 \cdot \hat{\mathbf{t}}. \tag{5.3.8}$$

In summary, the boundary conditions for the electromagnetic field at an interface between two planar media are such that

1. the in-plane components[6] of **E** and **H** are continuous across the interface.
2. the normal components[7] of **D** and **B** are continuous across the interface.

---

[6] i.e. the components that lie in a plane parallel to the plane of the interface, and therefore orthogonal to the plane of incidence.

[7] i.e. the components that are normal to the interface.

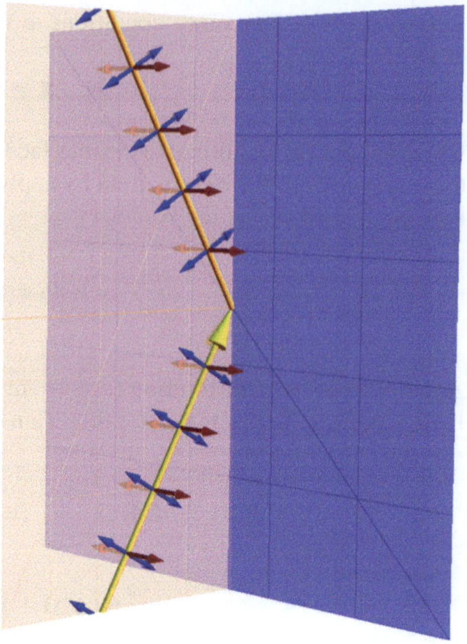

**Fig. 5.2** Reflection in the plane. The incident wave-vector $\mathbf{k}_1$ (the *yellow arrow*) strikes an interface and is partially reflected. The incident and reflected wave-vectors both lie in the plane of incidence. Two polarisations are depicted: the *blue arrows* depict the polarisation in which the electric field lies parallel to the plane of incidence; the *red arrows* depict the polarisation that lies orthogonal to it

#### 5.3.1.3 *s* and *p* Polarisations

We consider two polarisations of the electromagnetic field separately. Conventionally, we separate the field into one polarisation in which the waves have electric field orthogonal to the plane of incidence (the plane of the incident and reflected waves), and another in which the electric field lies parallel to it (i.e. the magnetic field is orthogonal to the plane of incidence). We refer to these as $s$ and $p$ polarisations respectively. Because of the linearity of Maxwell's equations, an arbitrary sum of the two polarisations is the solution for a general plane wave (Fig. 5.2).

#### 5.3.1.4 Boundary Conditions Applied to the Electric Field

Consider a single interface between two media: medium 1 $\{\epsilon_1, \mu_1, n_1 = \sqrt{\epsilon_1\mu_1}\}$ and medium 2 $\{\epsilon_2, \mu_2, n_2 = \sqrt{\epsilon_2\mu_2}\}$. Real media are dispersive,[8] so the permittivities and permeabilities are functions of the frequency $\omega$. We can write an expression for the field in medium 1, for a given frequency $\omega$, in terms of the $s$ and $p$ polarisations for waves propagating forwards and backwards:

$$\mathbf{E}_1 = E_{1s}^{(+)}(\mathbf{x})\mathbf{e}_{1s}^{(+)} + E_{1s}^{(-)}(\mathbf{x})\mathbf{e}_{1s}^{(-)} + E_{1p}^{(+)}(\mathbf{x})\mathbf{e}_{1p}^{(+)} + E_{1p}^{(-)}(\mathbf{x})\mathbf{e}_{1p}^{(-)}. \quad (5.3.9)$$

[8] Dissipation is incorporated by including an imaginary component in the refractive index.

The orientations of the $s$ and $p$ waves are given by the unit vectors

$$\mathbf{e}_{1s}^{(\pm)} = \hat{\mathbf{x}} \times \hat{\mathbf{k}}_{\parallel}, \tag{5.3.10a}$$

$$\mathbf{e}_{1p}^{(\pm)} = \hat{\mathbf{k}_1}^{(\pm)} \times \mathbf{e}_{1s}^{(\pm)}, \tag{5.3.10b}$$

where $\mathbf{k}_1^{(\pm)} = \pm k_{1x}\hat{\mathbf{x}} + \mathbf{k}_{\parallel}$, $\hat{\mathbf{x}}$ is a unit vector normal to the interface, $k_{1x}$ is the component of the wave-vector normal to the interface, and $\hat{\mathbf{k}}_{\parallel} = \mathbf{k}_{\parallel}/k_{\parallel}$ defines a plane parallel to the interface.[9] Applying the continuity of the in-plane part of the electric field across the two interfaces,[10]

$$\mathbf{E}_1 \cdot \mathbf{e}_{1s}^{(\pm)} = \mathbf{E}_2 \cdot \mathbf{e}_{2s}^{(\pm)}, \tag{5.3.11}$$

$$\mathbf{E}_1 \cdot \hat{\mathbf{k}}_{\parallel} = \mathbf{E}_2 \cdot \hat{\mathbf{k}}_{\parallel}, \tag{5.3.12}$$

yields the equations

$$E_{1s}^{(+)} + E_{1s}^{(-)} = E_{2s}^{(+)} + E_{2s}^{(-)}, \tag{5.3.13}$$

$$\frac{k_{1x}}{n_1}\left[E_{1p}^{(+)} - E_{1p}^{(-)}\right] = \frac{k_{2x}}{n_2}\left[E_{2p}^{(+)} - E_{2p}^{(-)}\right], \tag{5.3.14}$$

having noted[11] that

$$\mathbf{e}_{1p}^{(\pm)} \cdot \mathbf{e}_{1s}^{(\pm)} = 0, \quad \hat{\mathbf{k}}_{\parallel} \cdot \mathbf{e}_{1s}^{(\pm)} = 0, \quad \hat{\mathbf{k}}_{\parallel} \cdot \mathbf{e}_{1p}^{(\pm)} = \mp\frac{ck_{1x}}{n_1\omega}. \tag{5.3.15}$$

Then, applying continuity for the normal component of the displacement field,

$$\mathbf{D}_1 \cdot \hat{\mathbf{x}} = \mathbf{D}_2 \cdot \hat{\mathbf{x}}, \tag{5.3.16}$$

we obtain

$$\frac{\epsilon_1}{n_1}\left[E_{1p}^{(+)} + E_{1p}^{(-)}\right] = \frac{\epsilon_2}{n_2}\left[E_{2p}^{(+)} + E_{2p}^{(-)}\right], \tag{5.3.17}$$

having noted[12] that

---

[9] By definition, a polarisation aligned with $\mathbf{e}_{1s}^{(\pm)}$ excludes any normal component that crosses the interface; the electric field is therefore orthogonal to the plane of incidence. By definition, a polarisation aligned with $\mathbf{e}_{1p}^{(\pm)}$ lies in an orthogonal plane to any polarisation aligned with $\mathbf{e}_{1s}^{(\pm)}$.

[10] This involves two projections, as there are two orthogonal polarisations that we wish to treat separately, each of which contains a component in the plane parallel to the interface.

[11] For the third identity:

$$\hat{\mathbf{k}}_{\parallel} \cdot \mathbf{e}_{1p}^{(\pm)} = \hat{\mathbf{k}}_{\parallel} \cdot \left(\hat{\mathbf{k}_1}^{(\pm)} \times \mathbf{e}_{1s}^{(\pm)}\right) = \frac{\mathbf{k}_{\parallel}}{k} \cdot \left\{(\pm k_{1x}\mathbf{x} + \mathbf{k}_{\parallel}) \times \left(\hat{\mathbf{x}} \times \hat{\mathbf{k}}_{\parallel}\right)\right\} = \mp\frac{k_{1x}}{k} = \mp\frac{ck_{1x}}{n_1\omega}.$$

[12] For this identity:

$$\hat{\mathbf{x}} \cdot \mathbf{e}_{1p}^{(\pm)} = \frac{ck_{\parallel}}{n_1 \omega}. \tag{5.3.18}$$

There is no similar contribution for the $s$ component, whose field is orthogonal to the plane of incidence, $\hat{\mathbf{x}} \cdot \mathbf{e}_{1s}^{(\pm)} = 0$, and therefore has no normal component across the interface.

#### 5.3.1.5 Boundary Conditions Applied to the Magnetic Field

We determine the magnetic field via

$$i\omega \mathbf{B}_1 = \nabla \times \mathbf{E}_1, \tag{5.3.19}$$

which implies

$$\mathbf{B}_1 = \frac{n_1}{c}\left[E_{1s}^{(+)}(\mathbf{x})\mathbf{e}_{1p}^{(+)} + E_{1s}^{(-)}(\mathbf{x})\mathbf{e}_{1p}^{(-)} - E_{1p}^{(+)}(\mathbf{x})\mathbf{e}_{1s}^{(+)} - E_{1p}^{(-)}(\mathbf{x})\mathbf{e}_{1s}^{(-)}\right]. \tag{5.3.20}$$

The continuity of $\mathbf{k}_{\parallel} \cdot \mathbf{H}$ gives the boundary condition

$$\frac{k_{1x}}{\mu_1}\left[E_{1s}^{(+)} - E_{1s}^{(-)}\right] = \frac{k_{2x}}{\mu_2}\left[E_{2s}^{(+)} - E_{2s}^{(-)}\right]. \tag{5.3.21}$$

Applying the remaining conditions on the magnetic field does not generate any new equations.

#### 5.3.1.6 The Component Transfer Matrices

From the four equations obtained by applying the boundary conditions, we can form the matrix equation

$$\begin{pmatrix} 1 & 1 & 0 & 0 \\ \frac{k_{1x}}{\mu_1} & -\frac{k_{1x}}{\mu_1} & 0 & 0 \\ 0 & 0 & \frac{\epsilon_1}{n_1} & \frac{\epsilon_1}{n_1} \\ 0 & 0 & \frac{k_{1x}}{n_1} & -\frac{k_{1x}}{n_1} \end{pmatrix} \begin{pmatrix} E_{1s}^{(+)} \\ E_{1s}^{(-)} \\ E_{1p}^{(+)} \\ E_{1p}^{(-)} \end{pmatrix} = \begin{pmatrix} 1 & 1 & 0 & 0 \\ \frac{k_{2x}}{\mu_2} & -\frac{k_{2x}}{\mu_2} & 0 & 0 \\ 0 & 0 & \frac{\epsilon_2}{n_2} & \frac{\epsilon_2}{n_2} \\ 0 & 0 & \frac{k_{2x}}{n_2} & -\frac{k_{2x}}{n_2} \end{pmatrix} \begin{pmatrix} E_{2s}^{(+)} \\ E_{2s}^{(-)} \\ E_{2p}^{(+)} \\ E_{2p}^{(-)} \end{pmatrix}. \tag{5.3.22}$$

Clearly, the $s$ and $p$ polarisations are not coupled and can be treated separately. For the $s$ polarisation

---

$$\hat{\mathbf{x}} \cdot \mathbf{e}_{1p}^{(\pm)} = \frac{1}{k_1}\hat{\mathbf{x}} \cdot \left\{\pm k_{1x}\mathbf{x} \times \mathbf{e}_{1s}^{(\pm)} + \mathbf{k}_{\parallel} \times \left(\hat{\mathbf{x}} \times \hat{\mathbf{k}_{\parallel}}\right)\right\} = \frac{k_{\parallel}}{k_1}\hat{\mathbf{x}} \cdot \left\{\hat{\mathbf{k}_{\parallel}} \times \left(\hat{\mathbf{x}} \times \hat{\mathbf{k}_{\parallel}}\right)\right\} = \frac{ck_{\parallel}}{n_1\omega}.$$

$$\begin{pmatrix} 1 & 1 \\ \frac{k_{1x}}{\mu_1} & -\frac{k_{1x}}{\mu_1} \end{pmatrix}\begin{pmatrix} E_{1s}^{(+)} \\ E_{1s}^{(-)} \end{pmatrix} = \begin{pmatrix} 1 & 1 \\ \frac{k_{2x}}{\mu_2} & -\frac{k_{2x}}{\mu_2} \end{pmatrix}\begin{pmatrix} E_{2s}^{(+)} \\ E_{2s}^{(-)} \end{pmatrix}. \tag{5.3.23}$$

Rewriting this expression, to express the field quantities on the right in terms of the field quantities on the left, we obtain:

$$\begin{pmatrix} E_{2s}^{(+)} \\ E_{2s}^{(-)} \end{pmatrix} = \frac{\mu_2}{2k_{2x}}\begin{pmatrix} \frac{k_{2x}}{\mu_2} + \frac{k_{1x}}{\mu_1} & \frac{k_{2x}}{\mu_2} - \frac{k_{1x}}{\mu_1} \\ \frac{k_{2x}}{\mu_2} - \frac{k_{1x}}{\mu_1} & \frac{k_{2x}}{\mu_2} + \frac{k_{1x}}{\mu_1} \end{pmatrix}\begin{pmatrix} E_{1s}^{(+)} \\ E_{1s}^{(-)} \end{pmatrix}.$$

Similarly, for the $p$ polarisation:

$$\begin{pmatrix} E_{2p}^{(+)} \\ E_{2p}^{(-)} \end{pmatrix} = \frac{n_2}{2n_1\epsilon_2 k_{2x}}\begin{pmatrix} \epsilon_1 k_{2x} + \epsilon_2 k_{1x} & \epsilon_1 k_{2x} - \epsilon_2 k_{1x} \\ \epsilon_1 k_{2x} - \epsilon_2 k_{1x} & \epsilon_1 k_{2x} + \epsilon_2 k_{1x} \end{pmatrix}\begin{pmatrix} E_{1p}^{(+)} \\ E_{1p}^{(-)} \end{pmatrix}.$$

Thus for an arbitrary interface, composed of a homogeneous medium on one side with dielectric properties $\epsilon_l$ and $\mu_l$, and a second homogeneous medium on the other side with properties $\epsilon_r$ and $\mu_r$, it follows that the $s$ and $p$ transfer matrices connecting the field across the boundary between the two media are

$$\mathbf{t}_s(l,r) = \frac{1}{2\mu_l k_{xr}}\begin{pmatrix} \mu_l k_{xr} + \mu_r k_{xl} & \mu_l k_{xr} - \mu_r k_{xl} \\ \mu_l k_{xr} - \mu_r k_{xl} & \mu_l k_{xr} + \mu_r k_{xl} \end{pmatrix}, \tag{5.3.24}$$

$$\mathbf{t}_p(l,r) = \frac{n_r}{2n_l\epsilon_r k_{xr}}\begin{pmatrix} \epsilon_l k_{xr} + \epsilon_r k_{xl} & \epsilon_l k_{xr} - \epsilon_r k_{xl} \\ \epsilon_l k_{xr} - \epsilon_2 k_{xl} & \epsilon_l k_{xr} + \epsilon_r k_{xl} \end{pmatrix}. \tag{5.3.25}$$

We will be considering a stack of homogeneous media of equal width $a$, indexed by the parameter $m$. Consequently, we also require the transfer matrix associated with the propagation of a field through a homogeneous slice, in which the field accumulates a phase of $e^{ik_{xm}a}$:

$$\Lambda(m) = \begin{pmatrix} e^{ik_{xm}a} & 0 \\ 0 & e^{-ik_{xm}a} \end{pmatrix}. \tag{5.3.26}$$

#### 5.3.1.7 Imaginary Frequencies

In order to compute the Casimir stress or force integrals, we work in imaginary frequencies: $\omega \to ic\kappa$. This transforms the wave vector:

$$\omega^2 = \frac{c^2}{n^2}k^2 = \frac{c^2}{n^2}\left(k_x^2 + k_\parallel^2\right) \implies k_x = \sqrt{\frac{n^2\omega^2}{c^2} - k_\parallel^2} \to i\sqrt{n^2\kappa^2 + k_\parallel^2}. \tag{5.3.27}$$

We then denote the imaginary wave number

$$w_m = \sqrt{n_m^2 \kappa^2 + k_\parallel^2}. \tag{5.3.28}$$

At imaginary frequency, the transfer matrices are

$$\mathbf{t}_s(l, r) = \frac{1}{2\mu_l w_r} \begin{pmatrix} \mu_l w_r + \mu_r w_l & \mu_l w_r - \mu_r w_l \\ \mu_l w_r - \mu_r w_l & \mu_l w_r + \mu_r w_l \end{pmatrix}, \tag{5.3.29a}$$

$$\mathbf{t}_p(l, r) = \frac{n_r}{2n_l \epsilon_r w_r} \begin{pmatrix} \epsilon_l w_r + \epsilon_r w_l & \epsilon_l w_r - \epsilon_r w_l \\ \epsilon_l w_r - \epsilon_r w_l & \epsilon_l w_r + \epsilon_r w_l \end{pmatrix}. \tag{5.3.29b}$$

The permittivities $\epsilon$ and permeabilities $\mu$ are now evaluated at imaginary frequencies:

$$\epsilon(\omega) \to \epsilon(i\kappa), \quad \mu(\omega) \to \mu(i\kappa), \quad n(\omega) \to n(i\kappa). \tag{5.3.30}$$

These quantities are obtained from the dielectric properties for real frequencies by Hilbert transformation, but remain real-valued on the imaginary axis. The propagation matrix transforms to

$$\Lambda(m) = \begin{pmatrix} e^{-w_m a} & 0 \\ 0 & e^{w_m a} \end{pmatrix}, \tag{5.3.31}$$

which is also real.

### 5.3.2 Multilayer Transfer Matrix and Reflection Coefficients

Our intention is to track the properties of the electromagnetic field through an inhomogeneous medium, by modelling the medium as a stack of homogeneous slices. To this end, we seek to construct 'multilayer transfer matrices' that track wave propagation across multiple interfaces:

$$\mathbf{T}_\lambda(l, r) = \prod_{m=l}^{r-1} \Lambda(m+1)\mathbf{t}_\lambda(m, m+1). \tag{5.3.32}$$

This object connects the field emerging from the far right-hand side of a stack of homogeneous slices to the incident field impinging upon the stack in slice $l$:

$$\begin{pmatrix} E_{r\lambda}^{(+)} \\ E_{r\lambda}^{(-)} \end{pmatrix} = \mathbf{T}_\lambda(l, r) \begin{pmatrix} E_{l\lambda}^{(+)} \\ E_{l\lambda}^{(-)} \end{pmatrix} = \begin{pmatrix} T_{11} & T_{12} \\ T_{21} & T_{22} \end{pmatrix} \begin{pmatrix} E_{l\lambda}^{(+)} \\ E_{l\lambda}^{(-)} \end{pmatrix}. \tag{5.3.33}$$

We can now define the reflection coefficients for an interface composed of multiple slices. If we imagine a wave of unit amplitude incident from the left onto a planar object represented by $\mathbf{T}$, then the field emerging on the far right, into empty or homogeneous space, will consist only of right-propagating waves. It follows that:

$$\begin{pmatrix} T_{11} & T_{12} \\ T_{21} & T_{22} \end{pmatrix} \begin{pmatrix} 1 \\ r_r \end{pmatrix} = \begin{pmatrix} t_r \\ 0 \end{pmatrix}, \tag{5.3.34}$$

where $t_r$ is the amplitude of the wave emerging on the right-hand side of this structure, and $r_r$ is the amplitude of the wave reflected back. This reflection coefficient is completely determined by the details of the structure contained in the total transfer matrix:

$$T_{21} + T_{22}r_r = 0 \implies r_r = -\frac{T_{21}}{T_{22}}. \tag{5.3.35}$$

Similarly, for a wave incident from the right:

$$\begin{pmatrix} T_{11} & T_{12} \\ T_{21} & T_{22} \end{pmatrix} \begin{pmatrix} 0 \\ t_l \end{pmatrix} = \begin{pmatrix} r_l \\ 1 \end{pmatrix}. \tag{5.3.36}$$

The reflection coefficient is again determined completely by the total transfer matrix.

$$T_{22}t_l = 1 \implies t_l = 1/T_{22} \implies r_l = \frac{T_{12}}{T_{22}}. \tag{5.3.37}$$

## 5.3.3 *Approximate Transfer Matrices for an Inhomogeneous Medium*

### 5.3.3.1 Restricted Regime

The Casimir stress within a multilayer dielectric stack, for a given slice of the structure, may be expressed in terms of its reflection coefficients (5.2.1). However, numerical computations[13] suggest that the stress of an inhomogeneous structure, modelled as a stack of homogeneous slices, diverges as the number of slices increases (i.e. as we approach the continuum limit, in which the slicing becomes infinitesimally thin). The suspicion is that the divergence takes place in the integral over $k_{\parallel}$. In order to get an analytic fix on this divergence, we restrict our attention to the regime in which the in-plane wave vector is large in comparison to the refractive index multiplied by the frequency, i.e. where $n\kappa/k_{\parallel} \ll 1$. This extends from some finite value of $k_{\parallel}$ to infinity. Under these conditions:

$$w_m = \sqrt{n^2\kappa^2 + k_{\parallel}^2} = k_{\parallel}\sqrt{1 + \frac{n^2\kappa^2}{k_{\parallel}^2}} = k_{\parallel}\left(1 + \frac{1}{2}\frac{n^2\kappa^2}{k_{\parallel}^2} + \cdots\right) \sim k_{\parallel}, \tag{5.3.38}$$

[13] See Sect. 5.4.2.

i.e. the quantity $w_m$ becomes constant throughout the structure. The transfer matrices (5.3.29a, 5.3.29b) can also be simplified:

$$\mathbf{t}_s(l,r) \sim \frac{1}{2\mu_l}\begin{pmatrix} \mu_l+\mu_r & \mu_l-\mu_r \\ \mu_l-\mu_r & \mu_l+\mu_r \end{pmatrix}, \tag{5.3.39a}$$

$$\mathbf{t}_p(l,r) \sim \frac{n_r}{2n_l\epsilon_r}\begin{pmatrix} \epsilon_l+\epsilon_r & \epsilon_l-\epsilon_r \\ \epsilon_l-\epsilon_r & \epsilon_l+\epsilon_r \end{pmatrix}. \tag{5.3.39b}$$

These expressions become increasingly exact as $k_\parallel$ increases. We shall refer to this approximation as the 'high wavenumber regime' (*hw*-regime). Consider an inhomogeneous medium situated in the region $x \in [0, L]$. To describe the Casimir stress in this medium, we will slice it into $N$ portions of width $a$, with $Na = L$, indexed by $m$. The slices $m = 0$ and $m = N + 1$ contain free space. Restricting ourselves to the *hw*-regime, we first rewrite the transfer matrices (5.3.39a, 5.3.39b):

$$\mathbf{t}_s(m,m+1) = \begin{pmatrix} 1+\frac{\Delta\mu_m}{2\mu_m} & -\frac{\Delta\mu_m}{2\mu_m} \\ -\frac{\Delta\mu_m}{2\mu_m} & 1+\frac{\Delta\mu_m}{2\mu_m} \end{pmatrix}, \tag{5.3.40a}$$

$$\mathbf{t}_p(m,m+1) = \frac{n_{m+1}\epsilon_m}{n_m\epsilon_{m+1}}\begin{pmatrix} 1+\frac{\Delta\epsilon_m}{2\epsilon_m} & -\frac{\Delta\epsilon_m}{2\epsilon_m} \\ -\frac{\Delta\epsilon_m}{2\epsilon_m} & 1+\frac{\Delta\epsilon_m}{2\epsilon_m} \end{pmatrix}, \tag{5.3.40b}$$

where

$$\Delta\mu_m = \mu_{m+1} - \mu_m, \quad \Delta\epsilon_m = \epsilon_{m+1} - \epsilon_m. \tag{5.3.41}$$

For propagation between two slices, we first cross a boundary, then propagate a distance $a$. We may define a composite transfer matrix for this process, beginning with the $s$ polarisation:

$$\tau_s(m+1) = \Lambda(m+1)\,\mathbf{t}_s(m,m+1) = \begin{pmatrix} e^{-k_\parallel a} & 0 \\ 0 & e^{k_\parallel a} \end{pmatrix}\begin{pmatrix} 1+\frac{\Delta\mu_m}{2\mu_m} & -\frac{\Delta\mu_m}{2\mu_m} \\ -\frac{\Delta\mu_m}{2\mu_m} & 1+\frac{\Delta\mu_m}{2\mu_m} \end{pmatrix}.$$

Multiplying the two matrices, and separating the terms, the composite transfer matrix can be cast in the form

$$\begin{pmatrix} \left\{1+\frac{\Delta\mu_m}{2\mu_m}\right\}e^{-k_\parallel a} & -\frac{\Delta\mu_m}{2\mu_m}e^{-k_\parallel a} \\ -\frac{\Delta\mu_m}{2\mu_m}e^{k_\parallel a} & \left\{1+\frac{\Delta\mu_m}{2\mu_m}\right\}e^{k_\parallel a} \end{pmatrix} = \begin{pmatrix} e^{-k_\parallel a} & 0 \\ 0 & e^{k_\parallel a} \end{pmatrix} + \frac{\Delta\mu_m}{2\mu_m}\begin{pmatrix} e^{-k_\parallel a} & -e^{-k_\parallel a} \\ -e^{k_\parallel a} & e^{k_\parallel a} \end{pmatrix}. \tag{5.3.42}$$

Proceeding similarly for the $p$ polarised waves, the composite transfer matrices for the two polarisations can be conveniently rewritten in the form

$$\tau_s(m+1) = \boldsymbol{\alpha} + \frac{\Delta\mu_m}{2\mu_m}\boldsymbol{\beta},$$
$$\tau_p(m+1) = \frac{n_{m+1}\epsilon_m}{n_m\epsilon_{m+1}}\left(\boldsymbol{\alpha} + \frac{\Delta\epsilon_m}{2\epsilon_m}\boldsymbol{\beta}\right), \tag{5.3.43}$$

where

$$\boldsymbol{\alpha} = \begin{pmatrix} e^{-k_\parallel a} & 0 \\ 0 & e^{k_\parallel a} \end{pmatrix}, \quad \boldsymbol{\beta} = \begin{pmatrix} e^{-k_\parallel a} & -e^{-k_\parallel a} \\ -e^{k_\parallel a} & e^{k_\parallel a} \end{pmatrix}. \tag{5.3.44}$$

#### 5.3.3.2 Approximating the Transfer Matrices

Besides working within the *hw*-regime, we may simplify the algebra further by limiting our enquiry to the general case of a weakly inhomogeneous medium. The optical profile of the medium under consideration remains arbitrary; we simply restrict ourselves either to the case of an inhomogeneity that is small in magnitude but fairly rapid in variation, or fairly large in magnitude but only slowly varying. Reflections are the result of inhomogeneities. In the case under consideration, we expect the reflection coefficients and the Casimir stress to be small. If there is a divergence in the Casimir stress for even a weakly varying (that is, a weakly reflective) medium, there is every reason to expect a divergence in a strongly varying (strongly reflective) medium.

First, to calculate the stress in a cell $l$, we require the reflection coefficients of the cell boundaries, $r_l$ and $r_r$, and therefore the transfer matrices associated with propagation of waves from cell $l$ throughout the multilayer stack, travelling to the left and to the right. We shall consider a configuration for $N+1$ cells. Consider the left transfer matrix for the $s$ polarisation:

$$\begin{aligned}\mathbf{T}_{Ls} &= \prod_{m=1}^{l} \tau_s(m) = \prod_{m=1}^{l}\left(\boldsymbol{\alpha} + \frac{\Delta\mu_{m-1}}{2\mu_{m-1}}\boldsymbol{\beta}\right) \\ &= \left(\boldsymbol{\alpha} + \frac{\Delta\mu_{l-1}}{2\mu_{l-1}}\boldsymbol{\beta}\right)\left(\boldsymbol{\alpha} + \frac{\Delta\mu_{l-2}}{2\mu_{l-2}}\boldsymbol{\beta}\right)\left(\boldsymbol{\alpha} + \frac{\Delta\mu_{l-3}}{2\mu_{l-3}}\boldsymbol{\beta}\right) + \cdots\end{aligned} \tag{5.3.45}$$

It is not possible to analytically evaluate (5.3.45) unless we make the second approximation we alluded to: it is equivalent to the Born approximation in quantum mechanics, where we assume that scattering is weak [15]. In electromagnetism, this means that the properties of the medium must change slowly as a function of position. Products of the transfer matrices can then be truncated to first order in $\Delta\epsilon$ and $\Delta\mu$ [14]. This approximation is quite well suited to our case, for it is the case where the value of the stress ought to be minimal. Thus for a weakly inhomogeneous medium we may neglect terms that are higher than first-order in $\Delta\mu_i$ and $\Delta\epsilon_i$. It follows that

$$\mathbf{T}_{Ls} \sim \boldsymbol{\alpha}^l + \frac{\Delta\mu_{l-1}}{2\mu_{l-1}}\boldsymbol{\beta}\boldsymbol{\alpha}^{l-1} + \frac{\Delta\mu_{l-2}}{2\mu_{l-2}}\boldsymbol{\alpha}\boldsymbol{\beta}\boldsymbol{\alpha}^{l-2} + \frac{\Delta\mu_{l-3}}{2\mu_{l-3}}\boldsymbol{\alpha}^2\boldsymbol{\beta}\boldsymbol{\alpha}^{l-3} + \cdots$$
$$= \boldsymbol{\alpha}^l + \sum_{m=1}^{l} \frac{\Delta\mu_{l-m}}{2\mu_{l-m}}\boldsymbol{\alpha}^{m-1}\boldsymbol{\beta}\boldsymbol{\alpha}^{l-m}. \tag{5.3.46}$$

It is convenient to reverse and reindex the sum:

$$\mathbf{T}_{Ls} \sim \boldsymbol{\alpha}^l + \sum_{m=0}^{l-1} \frac{\Delta\mu_m}{2\mu_m}\boldsymbol{\alpha}^{l-m-1}\boldsymbol{\beta}\boldsymbol{\alpha}^m. \tag{5.3.47}$$

The matrices in the final lines are then

$$\boldsymbol{\alpha}^l = \begin{pmatrix} e^{-k_\parallel la} & 0 \\ 0 & e^{k_\parallel la} \end{pmatrix}, \tag{5.3.48a}$$

$$\boldsymbol{\alpha}^{l-m-1}\boldsymbol{\beta}\boldsymbol{\alpha}^m = \begin{pmatrix} e^{-k_\parallel la} & -e^{k_\parallel (2m-l)a} \\ -e^{k_\parallel (l-2m)a} & e^{k_\parallel la} \end{pmatrix}. \tag{5.3.48b}$$

Thus the transfer matrix can be written

$$\mathbf{T}_{Ls} = \begin{pmatrix} \left[1 + \sum_{m=0}^{l-1} \frac{\Delta\mu_m}{2\mu_m}\right] e^{-k_\parallel la} & -\sum_{m=0}^{l-1} \frac{\Delta\mu_m}{2\mu_m} e^{-k_\parallel (l-2m)a} \\ -\sum_{m=0}^{l-1} \frac{\Delta\mu_m}{2\mu_m} e^{k_\parallel (l-2m)a} & \left[1 + \sum_{m=0}^{l-1} \frac{\Delta\mu_m}{2\mu_m}\right] e^{k_\parallel la} \end{pmatrix}. \tag{5.3.49}$$

For the transfer matrix associated with propagation through the right-hand side of the structure:

$$\mathbf{T}_{Rs} = \prod_{m=l+1}^{N+1} \tau_s(m) = \prod_{m=l+1}^{N+1} \left(\boldsymbol{\alpha} + \frac{\Delta\mu_{m-1}}{2\mu_{m-1}}\boldsymbol{\beta}\right)$$
$$= \left(\boldsymbol{\alpha} + \frac{\Delta\mu_N}{2\mu_N}\boldsymbol{\beta}\right)\left(\boldsymbol{\alpha} + \frac{\Delta\mu_{N-1}}{2\mu_{N-1}}\boldsymbol{\beta}\right)\left(\boldsymbol{\alpha} + \frac{\Delta\mu_{N-2}}{2\mu_{N-2}}\boldsymbol{\beta}\right) + \cdots \tag{5.3.50}$$

We apply the same approximation:

$$\mathbf{T}_{Rs} \sim \boldsymbol{\alpha}^{N+1-l} + \frac{\Delta\mu_N}{2\mu_N}\boldsymbol{\beta}\boldsymbol{\alpha}^{N-l} + \frac{\Delta\mu_{N-1}}{2\mu_{N-1}}\boldsymbol{\alpha}\boldsymbol{\beta}\boldsymbol{\alpha}^{N-l-1} + \frac{\Delta\mu_{N-2}}{2\mu_{N-2}}\boldsymbol{\alpha}^2\boldsymbol{\beta}\boldsymbol{\alpha}^{N-l-2} + \cdots$$
$$= \boldsymbol{\alpha}^{N+1-l} + \sum_{m=l}^{N} \frac{\Delta\mu_m}{2\mu_m}\boldsymbol{\alpha}^{N-m}\boldsymbol{\beta}\boldsymbol{\alpha}^{m-l}. \tag{5.3.51}$$

The matrices in the final lines are

$$\boldsymbol{\alpha}^{N+1-l} = \begin{pmatrix} e^{-k_\parallel (N+1-l)a} & 0 \\ 0 & e^{k_\parallel (N+1-l)a} \end{pmatrix}, \tag{5.3.52a}$$

$$\boldsymbol{\alpha}^{N-m}\boldsymbol{\beta}\boldsymbol{\alpha}^{m-l} = \begin{pmatrix} e^{-k_\parallel (N-l+1)a} & -e^{-k_\parallel (N+l+1-2m)a} \\ -e^{-k_\parallel (N+l+1-2m)a} & e^{k_\parallel (N-l+1)a} \end{pmatrix}. \tag{5.3.52b}$$

Thus the right-transfer matrix for $s$ polarised light is

$$\mathbf{T}_{Rs} = \begin{pmatrix} \left[1+\sum_{m=1}^{N}\frac{\Delta\mu_m}{2\mu_m}\right] e^{-k_\parallel (N-l+1)a} & -\sum_{m=1}^{N}\frac{\Delta\mu_m}{2\mu_m} e^{-k_\parallel (N+l+1-2m)a} \\ -\sum_{m=1}^{N}\frac{\Delta\mu_m}{2\mu_m} e^{k_\parallel (N+l+1-2m)a} & \left[1+\sum_{m=1}^{N}\frac{\Delta\mu_m}{2\mu_m}\right] e^{k_\parallel (N-l+1)a} \end{pmatrix}. \tag{5.3.53}$$

The form of $\mathbf{T}_{Lp}$ and $\mathbf{T}_{Rp}$, for the $p$ polarisation, can be determined without further calculation from their $s$ counterparts by simply substituting permeability for permittivity and multiplying by the prefactor

$$\frac{n_{N+1}\epsilon_l}{n_l\epsilon_{N+1}}. \tag{5.3.54}$$

We can now state the reflection coefficients, applying the same ratios of the elements of the appropriate transfer matrix as before (5.3.35), (5.3.37). For reflection from the left of $s$ polarised light, we obtain the coefficient

$$r_{Ls}(l) = -\frac{\sum_{m=0}^{l-1}\frac{\Delta\mu_m}{2\mu_m} e^{-k_\parallel (l-2m)a}}{\left[1+\sum_{m=0}^{l-1}\frac{\Delta\mu_m}{2\mu_m}\right] e^{k_\parallel l a}} = -\frac{\sum_{m=0}^{l-1}\frac{\Delta\mu_m}{\mu_m} e^{-2k_\parallel (l-m)a}}{\left(2+\sum_{j=0}^{l-1}\frac{\Delta\mu_j}{\mu_j}\right)}. \tag{5.3.55}$$

For reflection from the right of $s$ polarised light, we obtain

$$r_{Rs}(l) = \frac{\sum_{m=1}^{N}\frac{\Delta\mu_m}{2\mu_m} e^{k_\parallel (N+l+1-2m)a}}{\left[1+\sum_{m=1}^{N}\frac{\Delta\mu_m}{2\mu_m}\right] e^{k_\parallel (N-l+1)a}} = \frac{\sum_{m=1}^{N}\frac{\Delta\mu_m}{\mu_m} e^{-2k_\parallel (m-l)a}}{\left[2+\sum_{m=1}^{N}\frac{\Delta\mu_m}{\mu_m}\right]}. \tag{5.3.56}$$

The reflection coefficients of the $p$ polarisation can be recovered by replacing the permeabilities with the permittivities. It is convenient to introduce into all the reflection coefficients an additional phase factor, such that we can associate the reflection coefficients for each slice with a point $x_l$ at the centre of the slice. The complete set of reflection coefficients are then given as follows: for reflection of $s$ and $p$ polarised light from the left,

$$r_{Ls}(x_l) = -\frac{\sum_{m=0}^{l-1} \frac{\Delta\mu_m}{\mu_m} e^{-2k_\parallel(l-m-1/2)a}}{\left(2+\sum_{m=0}^{l-1}\frac{\Delta\mu_m}{\mu_m}\right)},$$

$$r_{Lp}(x_l) = -\frac{\sum_{m=0}^{l-1} \frac{\Delta\epsilon_m}{\epsilon_m} e^{-2k_\parallel(l-m-1/2)a}}{\left(2+\sum_{m=0}^{l-1}\frac{\Delta\epsilon_m}{\epsilon_m}\right)}; \tag{5.3.57}$$

for reflection from the right,

$$r_{Rs}(x_l) = \frac{\sum_{m=l}^{N} \frac{\Delta\mu_m}{\mu_m} e^{-2k_\parallel(m-l+1/2)a}}{\left(2+\sum_{m=l}^{N}\frac{\Delta\mu_m}{\mu_m}\right)},$$

$$r_{Rp}(x_l) = \frac{\sum_{m=l}^{N} \frac{\Delta\epsilon_m}{\epsilon_m} e^{-2k_\parallel(m-l+1/2)a}}{\left(2+\sum_{m=l}^{N}\frac{\Delta\epsilon_m}{\epsilon_m}\right)}. \tag{5.3.58}$$

## 5.4 The Casimir Stress in an Inhomogeneous Medium

To recapitulate briefly, we have approximated the inhomogeneous medium as a series of $N$ homogeneous strips of width $a$ (see Fig. 5.1). We have employed the transfer matrix technique for our analysis of the field [13–16], with the field in strip $j+1$ being related to that in $j$ by

$$\boldsymbol{E}_q(j+1) = \boldsymbol{t}_q(j+1)\cdot\boldsymbol{E}_q(j), \tag{5.4.1}$$

where the index $q$ labels the polarization as in (5.2.1), and $\boldsymbol{t}_q(j+1)$ is the transfer matrix relating the field on the far right of slice $j$ to that on the far right of slice $j+1$. In (6.2.11) the electric field amplitude, $\boldsymbol{E}_q$ is written as a two element vector containing the right (+) and left (−) going parts,

$$\boldsymbol{E}_q(j) = \begin{pmatrix} E_q^{(+)}(j) \\ E_q^{(-)}(j) \end{pmatrix}. \tag{5.4.2}$$

We number the transfer matrices in (6.2.11) from 1 to $N+1$, with $\epsilon_0$ and $\epsilon_{N+1}$ equal to the vacuum permittivity, and $\mu_0$ and $\mu_{N+1}$ the vacuum permeability. In each of these slices $\boldsymbol{t}_q$ is given by the usual expression for the transfer matrix in piece–wise homogeneous media (e.g. [14, 16]). For the imaginary frequencies, $\omega/c = i\kappa$, encountered within (5.2.1) the $x$–directed wave–vector in the $j$th slice is also imaginary, $k_j = iw_j$, where, $w_j = (n_j^2\kappa^2 + k_\parallel^2)^{1/2}$. The limit of $N \to \infty$ and $a \to 0$ will be taken in the final step of the calculation.

Our object is to apply this formalism to determine whether the stress tensor (5.2.1) remains finite when the properties of the medium are represented by continuous functions of position. We suspect that it is not: a divergence of (5.2.1) is anticipated in the integral over $k_\parallel = |\boldsymbol{k}_\parallel|$. Physically—considering the allowed modes on the real frequency axis—we can picture this divergent contribution arising due to waves of high $k_\parallel$ undergoing reflections from the inhomogeneity of the medium. As $k_\parallel$ is increased, these waves contribute ever more to the local value of the stress tensor, when presumably in reality they should not be supported by the medium at all. For the purpose of identifying this anticipated divergence, we have restricted our attention to the *hw*-regime of the integrand in (5.2.1), where the in–plane wave–vector is large in comparison to the 'refractive index' multiplied by the 'frequency', $n_j\kappa/k_\parallel \ll 1$.

### 5.4.1 The High-Wavenumber Divergence

The integral for the Casimir stress, at a position $x_l$ within the multilayer structure, can now be stated in terms of the quantities we have derived:

$$\sigma_{xx}(x_l) = 2\hbar c \sum_{\lambda=s,p} \int_0^\infty \frac{\mathrm{d}\kappa}{2\pi} \int_{\mathbb{R}^2} \frac{\mathrm{d}^2\mathbf{k}_\parallel}{(2\pi)^2} w \frac{r_{L\lambda}(x_l) r_{R\lambda}(x_l) e^{-2aw}}{1 - r_{L\lambda}(x_l) r_{R\lambda}(x_l) e^{-2aw}}, \tag{5.4.3}$$

where

$$w = \sqrt{n(x_l, i\kappa)^2\kappa^2 + k_\parallel^2}, \quad k_\parallel = \left|\mathbf{k}_\parallel\right|, \tag{5.4.4}$$

and the reflection coefficients $r_{s\lambda}$ are defined by (5.3.57), (5.3.58). $n(i\kappa)$ is the refractive index of slice $l$ evaluated at imaginary frequencies. To examine the behaviour of the stress in the integral over $k_\parallel$ we evaluate the integrand above at fixed $\kappa$ (that is, for a fixed frequency) and therefore at $n(x_l) = n(x_l, i\kappa)$. A semi-infinite part of the integral over $k_\parallel$ is taken in the regime of high wave numbers $[K, \infty)$, where $w \sim k_\parallel$. The quantity we wish to analyse is

$$I = \sum_{\lambda=s,p} \int_K^\infty \frac{\mathrm{d}k_y}{2\pi} \int_K^\infty \frac{\mathrm{d}k_z}{2\pi} k_\parallel \frac{r_{L\lambda} r_{R\lambda} e^{-2k_\parallel a}}{1 - r_{L\lambda} r_{R\lambda} e^{-2k_\parallel a}}. \tag{5.4.5}$$

It is convenient to rewrite the integral in polar coordinates:

$$\sum_{\lambda=s,p} \int_0^{2\pi} \frac{\mathrm{d}\theta}{2\pi} \int_K^\infty \frac{k_\parallel^2 \mathrm{d}k_\parallel}{2\pi} \frac{r_{L\lambda} r_{R\lambda} e^{-2k_\parallel a}}{1 - r_{L\lambda} r_{R\lambda} e^{-2k_\parallel a}}. \tag{5.4.6}$$

If this quantity is divergent, then the stress integral as a whole is divergent. This is true regardless of the absorbing properties of the medium, or its frequency dispersion profile. We can expand the denominator in the expression above in a series of ascending powers of the reflection coefficients, on the assumption that $r_{L\lambda}r_{R\lambda}e^{-2k_{\parallel}a} < 1$, so that this sum converges for all $a$ and all $k_{\parallel}$:

$$r_{L\lambda}r_{R\lambda}e^{-2k_{\parallel}a}\left(1 - r_{L\lambda}r_{R\lambda}e^{-2k_{\parallel}a}\right)^{-1}$$
$$= r_{L\lambda}r_{R\lambda}e^{-2k_{\parallel}a}\left[1 + r_{L\lambda}r_{R\lambda}e^{-2k_{\parallel}a} + \left(r_{L\lambda}r_{R\lambda}e^{-2k_{\parallel}a}\right)^2 + \cdots\right], \tag{5.4.7}$$

which can be written more concisely as

$$\sum_{n=0}\left(r_{L\lambda}r_{R\lambda}e^{-2k_{\parallel}a}\right)^{n+1}. \tag{5.4.8}$$

We may now rewrite $I$ and introduce the quantity $I_{\lambda n}$:

$$I = \frac{1}{2\pi}\sum_{n=0}^{\infty}\sum_{\lambda=s,p} I_{\lambda n}, \quad I_{\lambda n} = \int_K^{\infty} k_{\parallel}^2 \mathrm{d}k_{\parallel}\,(r_{L\lambda}r_{R\lambda})^{n+1}\,e^{-2(n+1)k_{\parallel}a}. \tag{5.4.9}$$

Consider the quantity $I_{s0} = \int_K^{\infty} k_{\parallel}^2 \mathrm{d}k_{\parallel} r_{Ls}r_{Rs}e^{-2k_{\parallel}a}$. First, we insert the expressions for the reflection coefficients:

$$I_{s0} = -\int_K^{\infty} k_{\parallel}^2 \mathrm{d}k_{\parallel}\frac{\sum_{m=0}^{l-1}\sum_{m'=l}^{N}\frac{\Delta\mu_m\Delta\mu_{m'}}{\mu_m\mu_{m'}}e^{-2k_{\parallel}(m'-m+1)a}}{\left(2+\sum_{m=0}^{l-1}\frac{\Delta\mu_m}{\mu_m}\right)\left(2+\sum_{m'=l}^{N}\frac{\Delta\mu_{m'}}{\mu_{m'}}\right)}. \tag{5.4.10}$$

The integral over $k_{\parallel}$ can now be evaluated. It is an integral of the form[14]

$$\int x^2 e^{cx}\mathrm{d}x = \frac{\partial^2}{\partial c^2}\int e^{cx}\mathrm{d}x = \frac{\partial^2}{\partial c^2}\left[\frac{1}{c}e^{cx}\right] = e^{cx}\left(\frac{x^2}{c} - \frac{2x}{c^2} + \frac{2}{c^3}\right). \tag{5.4.11}$$

Putting $c = -2\left(m' - m\right)a$, and integrating over $k_{\parallel}$, we obtain

$$\int_K^{\infty} \mathrm{d}k_{\parallel}k_{\parallel}^2 e^{-2k_{\parallel}(m'-m+1)a} = K_0\,e^{-2K(m'-m+1)a} \tag{5.4.12}$$

where

$$K_0 = \left(\frac{K^2}{2(m'-m+1)a} + \frac{2K}{4(m'-m+1)^2a^2} + \frac{2}{8(m'-m+1)^3a^3}\right). \tag{5.4.13}$$

[14] We have set the constant of integration to zero.

Thus the integral $I_{s0}$ evaluates as

$$I_{s0} = -\frac{\sum_{m=0}^{l-1}\sum_{m'=l}^{N}\frac{\Delta\mu_m\Delta\mu_{m'}}{\mu_m\mu_{m'}}K_0\,e^{-2K(m'-m+1)a}}{\left(2+\sum_{m=0}^{l-1}\frac{\Delta\mu_m}{\mu_m}\right)\left(2+\sum_{m'=l}^{N}\frac{\Delta\mu_{m'}}{\mu_{m'}}\right)}. \tag{5.4.14}$$

In fact, this quantity can be made arbitrarily large by slicing the medium more finely—that is, by decreasing the width of each slice by increasing the number of slices. To see this, we consider the quantity $I_{s0}$ in the continuum limit. First, we rewrite the sums as integrals:

$$\sum_{m=0}^{l-1} \to \int_0^x \mathrm{d}x_1, \quad \sum_{m'=l}^{N} \to \int_x^L \mathrm{d}x_2. \tag{5.4.15}$$

The terms involving the permeabilities can be reexpressed as logarithms.

$$\frac{\Delta\mu_m}{\mu_m} = \frac{\mu_{m+1}-\mu_m}{\mu_m} \to \frac{1}{\mu(x_1)}\frac{\mathrm{d}}{\mathrm{d}x_1}\mu(x_1) = \frac{\mathrm{d}}{\mathrm{d}x_1}\ln\left[\mu(x_1)\right], \tag{5.4.16}$$

Similarly,

$$\frac{\Delta\mu_{m'}}{\mu_{m'}} \to \frac{\mathrm{d}}{\mathrm{d}x_2}\ln\left[\mu(x_2)\right]. \tag{5.4.17}$$

The sums on the denominator of (5.4.14) becomes integrals that can be straightforwardly evaluated:

$$\begin{aligned}\int_0^x \mathrm{d}x_1\frac{\mathrm{d}}{\mathrm{d}x_1}\ln\left[\mu(x_1)\right] &= \ln\left[\mu(x)\right]-\ln\left[\mu(0)\right],\\ \int_x^L \mathrm{d}x_2\frac{\mathrm{d}}{\mathrm{d}x_2}\ln\left[\mu(x_2)\right] &= \ln\left[\mu(L)\right]-\ln\left[\mu(x)\right].\end{aligned} \tag{5.4.18}$$

Since the medium is situated in a vacuum, $\ln[\mu(0)] = \ln[\mu(L)] = \ln 1 = 0$. Finally, we arrive at an expression for (5.4.14) in the continuum limit: $I_{s0}$ is equal to

$$-\left(2+\ln\left[\mu(x_1)\right]\right)^{-1}\left(2-\ln\left[\mu(x_2)\right]\right)^{-1}\int_0^x \mathrm{d}x_1\int_x^L \mathrm{d}x_2\frac{\mathrm{d}}{\mathrm{d}x_1}\ln\left[\mu(x_1)\right]\frac{\mathrm{d}}{\mathrm{d}x_2}\ln\left[\mu(x_2)\right]$$
$$\left(\frac{K^2}{2(x_2-x_1)a}+\frac{2K}{4(x_2-x_1)^2a^2}+\frac{2}{8(x_2-x_1)^3a^3}\right)e^{-2K(x_2-x_1)a}. \tag{5.4.19}$$

This expression clearly diverges. It therefore seems that there is no finite continuum limit of the regularised stress tensor (5.2.1). Including the additional terms in the series (5.4.9) will not affect this result [4]. Whilst these contributions diverge in a similar manner, they represent higher powers of the derivatives of $\epsilon$ and $\mu$—terms that vary quite independently as the spatial dependence of $\epsilon$ and $\mu$ is changed, and therefore cannot be expected to cancel in general. As the remainder of the integral over $\boldsymbol{k}_{\|}$ is finite, we conclude that the whole integral diverges as $a \to 0$. Consequently (5.2.1) diverges everywhere within an inhomogeneous medium described by $\epsilon$ and $\mu$ that are continuous functions of position. This is independent of how these quantities depend on imaginary frequency.

### 5.4.2 A Numerical Illustration

The divergence demonstrated analytically above was first spotted numerically when attempting to compute a stress profile for a system similar to the one in [9], and the results of these numerical computations serve to illustrate the argument. For the sake of simplicity we consider an impedance-matched system $\epsilon = \mu = n$ with the refractive index profile

$$n(x) = \begin{cases} 3 & x \le 0, \\ 3\,\mathrm{e}^{-x} & 0 < x < \mathrm{Log}\,3, \\ 1 & x \ge \mathrm{Log}\,3. \end{cases} \tag{5.4.20}$$

The system contains an imhomogeneous region between $x = 0$ and $x = \mathrm{Log}\,3$. In order to investigate the properties of this system using the transfer matrix technique described earlier (but dispensing with the approximations introduced in Sect. 5.3.3), we divide the inhomogeneous region into $N$ homogeneous pieces (see Fig. 5.3), and determine the left and right reflection coefficients within each piece. It is then possible

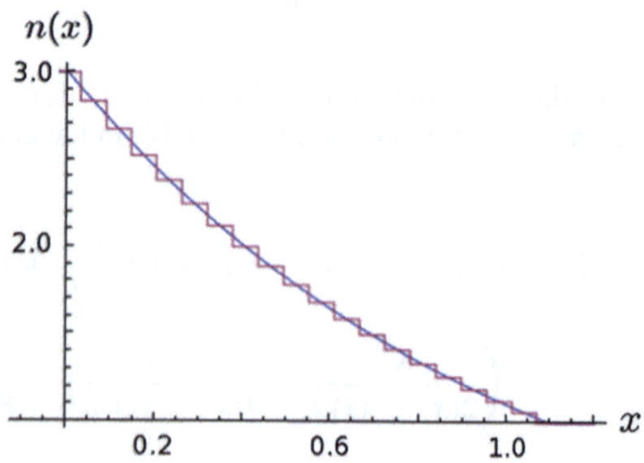


**Fig. 5.3** The continuous refractive index profile of the system, and a piece–wise approximation using 20 homogeneous slices

to calculate the local value of the regularised stress. The formula for the stress (5.2.1) can be rewritten more simply in this case [10], noting that the coefficients depend only on the magnitude of the wave vector components, and not on the angle between them[15]:

$$\sigma = \frac{\hbar c}{\pi^2} \int_0^\infty \int_0^\infty k_\parallel w \frac{r_L r_R \mathrm{e}^{-2aw}}{1 - r_L r_R \mathrm{e}^{-2aw}} dk_\parallel \, d\kappa. \tag{5.4.21}$$

As $N$ becomes large (i.e. as the cavity width $a$ becomes small), the approximation becomes increasingly accurate. *Prima facie*, there should be little to distinguish the physics of the case $N = 400$ from the case $N = 800$, as both approximations of the continuum case are now very smooth. Nevertheless, as Fig. 5.4 shows, the stress (though regularised) increases markedly, and it continues to grow as more slices are added. Why is this happening? Plots of the integrand of the stress (6.2.7), where the wave number $k_\parallel$ and the number of slices $N$ are allowed to vary, show that the integral falls off less and less rapidly with $k_\parallel$ as $N$ is increased (Fig. 5.5).

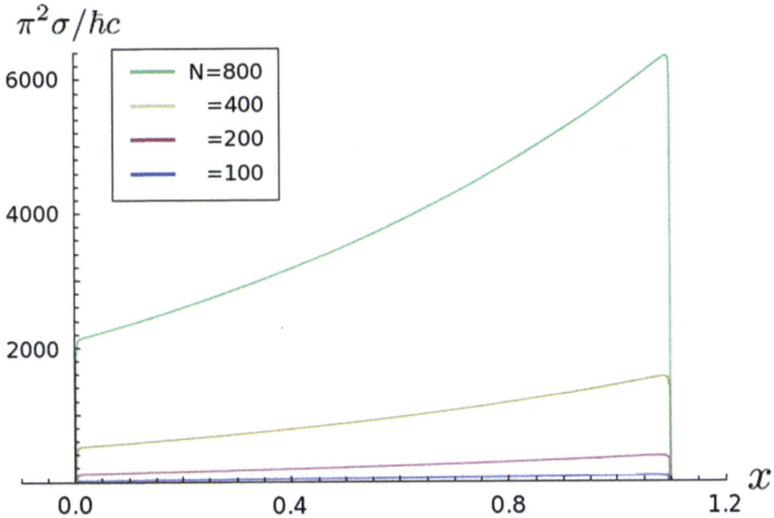


**Fig. 5.4** The medium, inhomogeneous between $x = 0$ and $x = \text{Log}\,3$, is divided into 100, 200, 400 and 800 homogeneous slices. The local absolute value of the regularised stress tensor (6.2.7)—normalised in units of $\hbar c/\pi^2$—is plotted for each case at a given position $x$. The stress increases as the number of divisions increases

[15] We also note that the electric and magnetic coefficients are equal, due to impedance-matching, and hence need not be referred to separately.

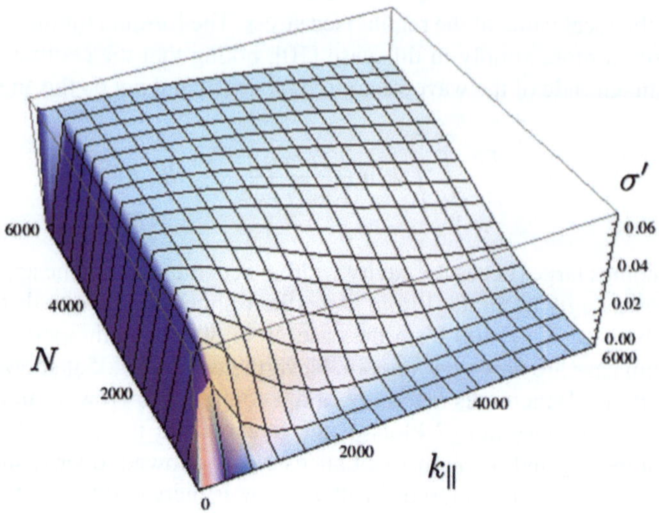


**Fig. 5.5** The integrand of the stress (6.2.7), $\sigma'$, (normalised in the same units as Fig. 5.4) is plotted for $\kappa = 1$ at the centre of the system, with $k_{\parallel}$ varying from $k_{\parallel} = 1$ to $k_{\parallel} = 6,000$ (horizontal axis), and $N$ ranging from $t = 10$ to $t = 6,000$ (depth axis). As the number of slices $N$ is increased, the integrand falls off less rapidly with $k_{\parallel}$, and thus the integral of the stress converges less rapidly

### *5.4.3 Speculations on Spatial Dispersion*

We might wonder how finite results ought to be extracted from this formalism. The advantage of the usual regularisation procedure is that it removes an infinite quantity that does not depend on the inhomogeneity of the medium, which cannot be relevant to the force. Conversely, here we have a divergent contribution that is due to the inhomogeneity of the medium. The divergence originates within the fact that the reflection coefficients (5.3.57), (5.3.58) do not go to zero fast enough as $k_{\parallel} \to \infty$ in the limit where $a \to 0$.

One approach to this problem might be to terminate the integral over $k_{\parallel}$ at some finite cut–off. However, the value of this cut–off would seem to be arbitrary. Alternatively, before the continuum limit is taken, we might just remove some small region of the sum around the point where $k - j = 1$, although the size of this region would also be arbitrary. This problem is reminiscent of that found in the case of spontaneous emission within an absorbing dielectric, where an additional physical parameter—equivalent to removing a portion of the dielectric in the immediate vicinity of the atom—must be introduced in order to obtain a finite emission rate [17, 18].

However, it might be urged that the solution to this problem involves the recognition that, as with other problems in physics, the specific dependence of the dielectric media on both the frequency *and* the wave vector ought to be included within the macroscopic description of matter, at the level of the electric permittivity and magnetic permeability functions. It is a familiar thought that realistic models of macroscopic media must include the phenomenon of *temporal* dispersion: in linear and causal time-independent systems, this amounts to the acknowledgement that the general relation between the electric displacement **D** and the electric field **E**, given by the dielectric response tensor of the medium, $\epsilon(\mathbf{r}, \mathbf{r}', t-t')$, involves the response of the system not only at time $t$ but also the delayed response to the field at previous times $t' < t$,

$$\mathbf{D}(\mathbf{r}, t) = \int \mathrm{d}\mathbf{r}' \int_{-\infty}^{t} \mathrm{d}t' \epsilon(\mathbf{r}, \mathbf{r}', t - t') \cdot \mathbf{E}(\mathbf{r}', t'). \tag{5.4.22}$$

The time integration can be rewritten as a temporal Fourier transform, and the response function replaced by the more familiar frequency-dependent electric permittivity, so that

$$\mathbf{D}(\mathbf{r}, t) = \int \mathrm{d}\mathbf{r}' \epsilon(\mathbf{r}, \mathbf{r}', \omega) \cdot \mathbf{E}(\mathbf{r}', \omega). \tag{5.4.23}$$

However, (5.4.22) also acknowledges that the response at a position **r** depends on positions within the locality of **r**, as well as the point of measurement itself. Strictly speaking, this *non-locality* holds for all materials at the microscopic level, as they are made of atoms, and the phenomenological approximation of a continuous medium is no longer applicable on this length scale. As the wavelength of light is typically much larger than the interatomic distance, it is generally reasonable to assume that $\mathbf{E}(\mathbf{r}') \approx \mathbf{E}(\mathbf{r})$, yielding a local response

$$\mathbf{D}(\mathbf{r}, \omega) = \epsilon(\mathbf{r}, \omega) \cdot \mathbf{E}(\mathbf{r}, \omega). \tag{5.4.24}$$

Nevertheless, it has been shown that non-local effects must be considered in order to model the behaviour of the Casimir force at finite temperature [19], and it has also been noted that this approximation may fail close to the surface of a material [20, 21]. For a system that is translationally invariant, with a response depending on the separation $\mathbf{r} - \mathbf{r}'$, we may write a new permittivity function by taking a spatial Fourier transform with the wave vector **k**, as well as a temporal Fourier transform with respect to time:

$$\epsilon(\mathbf{k}, \omega) = \int \mathrm{d}\mathbf{r}' \int_{-\infty}^{t} \mathrm{d}t' \epsilon(\mathbf{r} - \mathbf{r}', t - t') \cdot \mathbf{E}(\mathbf{r}', t'). \tag{5.4.25}$$

This non-local dependence is known as spatial dispersion, being formally similar to temporal dispersion. Properly, a complete model of dispersion requires the dielectric response of the medium to fall off to its free space (bare vacuum) magnitudes, for high values of frequency and wave number:

$$\lim_{k\to\infty} \epsilon(\mathbf{k}, \omega) = 1 \quad \text{and} \quad \lim_{\omega\to\infty} \epsilon(\mathbf{k}, \omega) = 1. \tag{5.4.26}$$

Importantly, if the response of the material falls off fast enough with high wave numbers, the stress tensor (5.2.1) may turn out to be finite after all. To examine this question, however, requires detailed empirical information about the material sample, or a sufficiently well-motivated theoretical model of dielectric, from which its high-wavenumber response can be modelled. Such probing questions into the detailed structure of macroscopic media are beyond the scope of this present study and seem unlikely to yield a fundamental solution: the Casimir energy of a system without spatial or temporal dispersion can be stated exactly, as we witnessed in the previous chapter, and the divergence of the stress does not arise from its temporally dispersive properties.

## 5.5 Summary Remarks

In this chapter we have examined the local behaviour of the regularised stress tensor commonly used in calculations of the Casimir force for a dielectric medium inhomogeneous in one direction. We have seen that the usual expression for the stress tensor is not finite anywhere within the medium, whatever the temporal dispersion or index profile, and that this divergence is unlikely to be removed by simply modifying the regularisation procedure. These findings hold for all magnetodielectric media.

From our investigation, it is clear that a calculation of (5.2.1) for a piecewise definition of an inhomogeneous medium does not represent an approximation to the continuous case. Our result is consistent with earlier findings,[16] and illustrates the generality of the problem of specifying the local value of the electromagnetic stress tensor at $T = 0\,\mathrm{K}$ when the material parameters vary continuously over space. Moreover, we identify a divergence of the local value of the stress tensor that cannot be removed by the procedure of regularisation usually advocated; it arises specifically due to the unphysical contribution of high wave numbers in the continuum limit. This problem does not seem to be widely appreciated in the literature. In [13] and in [22] reflection coefficients were similarly employed to determine the Casimir force in systems with increasingly refined inhomogeneous features, but the limits of the applicability of this technique were not commented on.

A continuously varying medium introduces arbitrarily small inhomogeneities into a system. A possible explanation for the divergence we have identified is that the Casmir force does not depend on such small-scale inhomogeneities, and a generally finite and physically meaningful result must be obtained by finding some simple modification to the existing regularisation procedure.

[16] In [9] the stress inside an inhomogeneous medium with a similar profile to (5.4.20) was determined using the exact Green function. However, it was infinite everywhere. An alternative regularisation was proposed, but this proved unsuccessful [4].

# References


1. I.E. Dzyaloshinskii, E.M. Lifshitz, L.P. Pitaevskii, Adv. Phys. **10**, 165 (1961)
2. J.N. Munday, F. Capasso, V.A. Parsegian, Nature **457**, 170 (2009)
3. A.W. Rodriguez, F. Capasso, S.G. Johnson, Nat. Photonics **5**, 211 (2011)
4. W.M.R. Simpson, S.A.R. Horsley, U. Leonhardt, Phys. Rev. A **87**, 043806 (2013)
5. T.G. Philbin, New J. Phys. **12**, 123008 (2010)
6. T.G. Philbin, New J. Phys. **13**, 063026 (2011)
7. L.D. Landau, E.M. Lifshitz, *Quantum Mechanics* (Butterworth-Heinemann, Oxford, 2005)
8. E. M. Lifshitz and L. P. Pitaevskii, *Statistical Physics* (Butterworth-Heinemann, Oxford, 2003)
9. T.G. Philbin, C. Xiong, U. Leonhardt, Ann. Phys. **325**, 579 (2009)
10. U. Leonhardt, *Essential Quantum Optics* (Cambridge University Press, Cambridge, 2010)
11. L. P. Pitaevskii, *Electrodynamics of Energy Loss Spectroscopy in the Electron Microscope* (Springer, New York, 2011)
12. E.M. Lifshitz, Zh. Eksp. Teor. Fiz. **29**, 94 (1955)
13. C. Genet, A. Lambrecht, S. Reynaud, Phys. Rev. A **67**, 043811 (2003)
14. M. Born, E. Wolf, *Principles of Optics* (Cambridge University Press, Cambridge, 1999)
15. I.A. Shelykh, V.K. Ivanov, Int. J. Theor. Phys. **43**, 477 (2004)
16. M. Artoni, G.C. La Rocca, F. Bassani, Phys. Rev. E **72**, 046604 (2005)
17. S.M. Barnett, B. Huttner, R. Loudon, R. Matloob, J. Phys. B: At. Mol. Opt. Phys. **29**, 3763 (1996)
18. S. Scheel, L. Knöll, D.-G. Welsch, Phys. Rev. A **60**, 4094 (1999)
19. B.E. Sernelius, Phys. Rev. B **71**, 235114 (2005)
20. R. Esquivel-Sirvent et al., J. Phys. A: Math. Gen. **39**, (6323) (2006)
21. V.B. Svetovoy, R. Esquivel, Rev. E, **72**, 036113 (2005)
22. S. Ellingsen, J. Phys. A: Math. Theor. **40**, 1951 (2007)

## References

1. [illegible]
2. [illegible] Science 327 [illegible] (2010)

# Part III
# Conundrums in Casimir Theory

# Chapter 6
# The Casimir Force in a 'Compressive' Medium

*The Tardis, when working properly, is capable of many amazing things, not unlike myself.*

The 6th Doctor

## 6.1 The Paradox of Transformation Media

As we have observed in the previous two chapters, it remains beyond the scope of present theory to predict both the nature and size of Casimir forces in many simple systems. For example, consider the case of a cavity with perfect mirrors—Casimir's original model. We know that it is possible to calculate a Casimir force using Lifshitz theory for an empty cavity [1], or a cavity filled with perfectly homogeneous fluid, such as purified water [2, 3]. A spoonful of sugar dissolved in the water, however, is enough to frustrate the calculation; the stress tensor yields an infinite force on the mirrors with even the gentlest perturbation in the optical properties of the medium, when the perturbation is described as a continuous function of position. It seems that Casimir forces are, in general, impossible to predict in inhomogeneous media using Lifshitz theory [4–6].

However, as an idealised thought-experiment, we can imagine introducing an inhomogeneous metamaterial into a cavity whose effect on light could be interpreted as a simple distortion of the laboratory coordinate system.[1] Such media are common to the field of transformation optics, and have been put to use in various applications (such as simple cloaking devices, for example) [8, 9]. It is inconceivable that such a modification to the cavity could alter the fundamental nature of the Casimir force.[2]

[1] The main results in this chapter were published in [7].

[2] Maxwell's equations in a transformed coordinate system (transformed from a Cartesian grid) in empty space take the same form as Maxwell's equations in a Cartesian system with a medium



W.M.R. Simpson, *Surprises in Theoretical Casimir Physics*, Springer Theses,
DOI 10.1007/978-3-319-09315-4_6

There may appear to be a contradiction, then, with our earlier conclusions: we have found that the stress tensor suffers a divergence when we attempt to calculate it within an inhomogeneous medium [4, 5]; however, as we will see, it is possible to imagine introducing an idealised inhomogeneous medium in the chamber that effectively modifies the size of an empty cavity—in which case, we can predict that the Casimir force should be finite.

In this chapter, we resolve this apparent contradiction. In so doing, we are able to determine an exact expression for the Casimir force when a certain inhomogeneous medium, which we will call the $C$-slice, is introduced between the walls of a reflective cavity. Analytic solutions for Casimir forces are extremely rare—the attraction between two plates with infinite electric permittivity [1], the repulsion between plates with infinite electric permittivity and magnetic permeability [10], and the attractive force on a homogeneous spherical ball with infinite permittivity have been solved [11]. There are perhaps only two exact solutions for the Casmir force in a gradually varying medium: the first is the subject of the next chapter; we will discuss the second one here.

## 6.2 The Casimir Force in a 'Compressed' Cavity

### 6.2.1 Properties of the $C$-Slice

Once again, we consider a simple modification of Casimir's cavity, in which a wafer characterised by anisotropic, inhomogeneous electric permittivity and magnetic permeability tensors

$$\epsilon = \mathrm{diag}(\epsilon_{xx}, \epsilon_{yy}, \epsilon_{zz}), \quad \mu = \mathrm{diag}(\mu_{xx}, \mu_{yy}, \mu_{zz}),$$

given by

$$\epsilon(z,\omega) = \mu(z,\omega) = \begin{pmatrix} m(z,\omega)^{-1} & 0 & 0 \\ 0 & m(z,\omega)^{-1} & 0 \\ 0 & 0 & m(z,\omega) \end{pmatrix} \tag{6.2.1}$$

is inserted between the perfect mirrors of the cavity (see Fig. 6.1). The cavity mirrors are positioned at $z = 0, d$, and the wafer occupies an interval $z \in [a, b] \subset [0, d]$. This interposing medium, which we will refer to as a $C$-slice,[3] is impedance-matched

(Footnote 2 continued)
present. These different pictures are referred to as the 'virtual' and the 'physical' space respectively [8]. In swapping between them, the physics should remain the same. Considered in the 'virtual space' of the effective geometry the device implements, the coordinate transformed values of the permittivity and permeability tensors remain homogeneous. In the coordinate system of physical space, however, the optical properties of the device vary continuously.

[3] The C here denotes a *compression* of the vacuum. The compressive properties of a homogeneous, non-dispersive form of this metamaterial have been discussed pedagogically in [12].

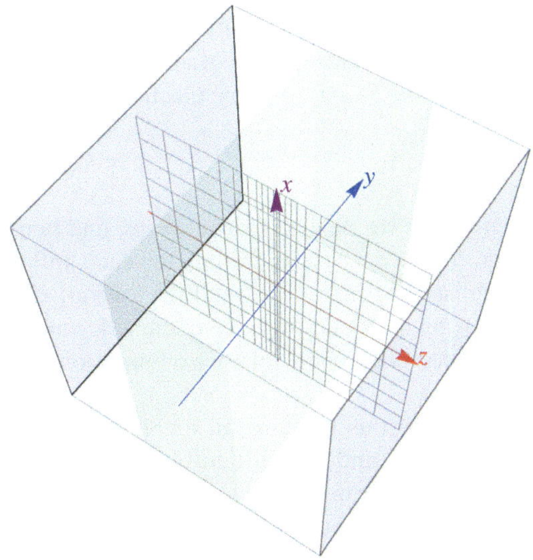


**Fig. 6.1** The $C$-slice (*green*) effectively compresses or expands the coordinate system of the cavity in one dimension along the direction in which its permittivity and permeability are varying inhomogeneously (the $z$-axis)

and intended to implement a coordinate transformation in which an interval of space is compressed along the $z$–axis by a factor

$$C_S = \frac{1}{\Delta} \int_a^b m(z, \omega)\, \mathrm{d}x, \quad \Delta = b - a. \tag{6.2.2}$$

It is not difficult to see why the $C$-slice should have these properties [12]. Beginning first with a homogeneous, non-dispersive $C$-slice, where the compression can be characterised by a simple factor $m(z, \omega) = const = m$, we consider an electromagnetic wave with wave number $k$ travelling parallel to the intended axis of compression. The wave must pick up the same phase, as it passes through the length of the device, as it would have acquired had it passed through an uncompressed region. Hence

$$k\Delta = km\Delta\sqrt{\epsilon_\perp \mu_\perp}. \tag{6.2.3}$$

Here, $\epsilon_\perp$ and $\mu_\perp$ are the permittivity and permeability components normal to the axis of compression; $\epsilon_\parallel$ and $\mu_\parallel$ are the components that are parallel to it. From (6.2.3) we deduce that

$$\epsilon_\perp \mu_\perp = m^{-2}. \tag{6.2.4}$$

In order to be equivalent to a distortion of the coordinate system, we require impedance-matched media, $\epsilon = \mu$, hence

$$\epsilon_\perp = \mu_\perp = m^{-1}. \tag{6.2.5}$$

Consider now a wave travelling along the $x$–axis, perpendicular to the axis of compression, and therefore free of any compression, polarised so that the magnetic field lies along the $z$–axis, and the electric field along the $y$–axis. In this case

$$k = k\sqrt{\epsilon_{\perp}\mu_{\parallel}}, \tag{6.2.6}$$

from which we deduce that $\mu_{\parallel} = m$, and hence $\epsilon_{\parallel} = m$. The generalisation to the inhomogeneous case is trivial: for each infinitesimal region $\delta z$, the space is now compressed to $m(z)^{-1}\delta z$, and the overall compression factor is given by (6.2.2). Finally, by incorporating a frequency dependence, we allow different frequencies to experience different levels of compression, and therefore allow for some degree of dispersion.

However, this remains an idealised model: the Casimir Effect is a broadband phenomenon, and it is difficult to see how the necessary condition of impedance-matching could in practice be secured for a sufficiently large portion of the electromagnetic spectrum.[4] Nevertheless, the practical difficulties of implementing this device should not disqualify its use in a *Gedankenexperiment*, and we will briefly discuss some possible applications for the $C$-slice later in this chapter.

## 6.2.2 Modifying the Casimir Force

### 6.2.2.1 Lifshitz Theory in Anisotropic Media

The vacuum stress of the cavity, according to Lifshitz theory, is determined by the scattering properties of its constituents, and can be codified in the form of reflection coefficients. In the case of anisotropic media, the form of the stress must be modified [14], replacing (5.2.1) with

$$\sigma_{xx}(x) = 2\hbar c \int_0^{\infty} \frac{d\kappa}{2\pi} \int_{\mathbb{R}^2} \frac{d^2\mathbf{k}_{\parallel}}{(2\pi)^2} w \,\mathrm{Tr}\left[\frac{\mathbf{R}_L\mathbf{R}_R e^{-2dw}}{1-\mathbf{R}_L\mathbf{R}_R e^{-2dw}}\right], \tag{6.2.7}$$

where the scalar reflection coefficients have been supplanted by matrices

$$\mathbf{R}_L = \begin{pmatrix} r_L^{ss} & r_L^{sp} \\ r_L^{ps} & r_L^{pp} \end{pmatrix}, \quad \mathbf{R}_R = \begin{pmatrix} r_R^{ss} & r_R^{sp} \\ r_R^{ps} & r_R^{pp} \end{pmatrix}. \tag{6.2.8}$$

Physically, this reformulation of the stress anticipates the phenomenon of conversion between polarisations that can take place in anisotropic media. $r_{L,R}^{pq}$ is the ratio of a field with $p$–polarization divided by an incoming field with $q$–polarization, for

[4] For a more positive appraisal of the utility of metamaterials for modifying the Casimir force, see [13].

reflection from the left (L) or right (R). The indices $s$ and $p$ correspond respectively to the customary perpendicular and parallel polarizations with respect to the plane of incidence. In the case of isotropic media, the off-diagonal elements of both matrices vanish, the diagonal elements are given by the familiar Fresnel expressions, and (6.2.7) reduces to the usual Lifshitz formula (5.2.1).

#### 6.2.2.2 4d Transfer Matrices

To determine how a $C$-slice modifies the Casimir force, we must compute the reflection coefficients of the device for all angles of incidence. The $C$-slice we are considering, in this case, is an inhomogeneous material; its optical properties vary continuously along the $z$−axis. A set of reflection coefficients may be obtained, however, by approximating the cavity as a series of $N+1$ homogeneous slices of width $\delta z$, where each slice has electric permittivity $\epsilon_n$ and magnetic permeability $\mu_n$, and $n \in [0, N+1]$ indexes the slicing. It is possible to compute such a set of coefficients for finite $N$, whether or not the stress tensor is divergent in the continuum limit (see Chap. 5). Between the plates, the electric and magnetic response is characterised by

$$\epsilon_n = \mu_n = \begin{pmatrix} m_n^{-1} & 0 & 0 \\ 0 & m_n^{-1} & 0 \\ 0 & 0 & m_n \end{pmatrix} \quad \text{for} \quad n \in [2, N]. \tag{6.2.9}$$

For an empty slice, $m_n = 1$. Once the reflection coefficients have been determined, the stress may be calculated. Adopting a slightly different notation from before, for convenience, we can write an expression for the field in a homogeneous slice in terms of two orthogonal polarisations, for waves propagating both forwards and backwards, $\lambda \in \{1, 2\}$ and $\lambda \in \{3, 4\}$:

$$\begin{pmatrix} \mathbf{E}_n \\ \mathbf{H}_n \end{pmatrix} = \sum_{\lambda=1}^{4} E_n^{(\lambda)} \begin{pmatrix} \mathbf{e}_n^{(\lambda)} \\ \mathbf{h}_n^{(\lambda)} \end{pmatrix}, \tag{6.2.10}$$

where $\mathbf{e}_n^{(\lambda)}$ and $\mathbf{h}_n^{(\lambda)}$ are the eigenmodes of the electric and magnetic fields respectively in a medium with the optical properties of slice $n$. There are four eigenmodes of the field to a slice, corresponding to two independent modes propagating forwards and backwards; the general solution for the field is a linear combination of all four of them. The transfer matrix technique [15–19] can again be used for such an analysis of the field, the field in strip $n+1$ being related to the field in $n$ by

$$\boldsymbol{E}(n+1) = \boldsymbol{t}(n+1) \cdot \boldsymbol{E}(n), \tag{6.2.11}$$

where $\boldsymbol{t}(n+1)$ is the $4 \times 4$ transfer matrix relating the field on the right side of a boundary between two media to the field on its left. As before, we obtain the transfer matrix by imposing the boundary conditions at the interface (see Sect. 5.3.1). We need only impose four of them to determine the field, as the electric and magnetic components are not independent of each other. For the convenience of this representation we will require that the tangential components of both fields should be continuous at the boundary:

$$\sum_\lambda E_n^{(\lambda)} \mathbf{e}_n^{(\lambda)} \cdot \hat{\mathbf{y}} = \sum_\lambda E_{n+1}^{(\lambda)} \mathbf{e}_{n+1}^{(\lambda)} \cdot \hat{\mathbf{y}}, \tag{6.2.12a}$$

$$\sum_\lambda E_n^{(\lambda)} \mathbf{h}_n^{(\lambda)} \cdot \hat{\mathbf{x}} = \sum_\lambda E_{n+1}^{(\lambda)} \mathbf{h}_{n+1}^{(\lambda)} \cdot \hat{\mathbf{x}}, \tag{6.2.12b}$$

$$\sum_\lambda E_n^{(\lambda)} \mathbf{h}_n^{(\lambda)} \cdot \hat{\mathbf{y}} = \sum_\lambda E_{n+1}^{(\lambda)} \mathbf{h}_{n+1}^{(\lambda)} \cdot \hat{\mathbf{y}}, \tag{6.2.12c}$$

$$\sum_\lambda E_n^{(\lambda)} \mathbf{e}_n^{(\lambda)} \cdot \hat{\mathbf{x}} = \sum_\lambda E_{n+1}^{(\lambda)} \mathbf{e}_{n+1}^{(\lambda)} \cdot \hat{\mathbf{x}}. \tag{6.2.12d}$$

Equation (6.2.12a–6.2.12d) can be written in the form of a $4d$ matrix equation:

$$\begin{pmatrix} E_{n+1}^{(1)} \\ E_{n+1}^{(2)} \\ E_{n+1}^{(3)} \\ E_{n+1}^{(4)} \end{pmatrix} = \mathbf{M}(n+1) \begin{pmatrix} E_n^{(1)} \\ E_n^{(2)} \\ E_n^{(3)} \\ E_n^{(4)} \end{pmatrix}. \tag{6.2.13}$$

The matrix $\mathbf{M}(n+1)$ implements the boundary conditions at the interface between slices $n$ and $n+1$,

$$\mathbf{M}(n+1) = \mathbf{D}(n+1)^{-1}\mathbf{D}(n), \tag{6.2.14}$$

and is computed for a system of two half-spaces. The matrix $\mathbf{D}(n)$ characterises the state of the electromagnetic field in a medium with the dielectric properties of slice $n$, and is defined

$$\mathbf{D}(n) = \begin{pmatrix} \mathbf{e}_n^{(1)} \cdot \hat{\mathbf{y}} & \mathbf{e}_n^{(2)} \cdot \hat{\mathbf{y}} & \mathbf{e}_n^{(3)} \cdot \hat{\mathbf{y}} & \mathbf{e}_n^{(4)} \cdot \hat{\mathbf{y}} \\ \mathbf{h}_n^{(1)} \cdot \hat{\mathbf{x}} & \mathbf{h}_n^{(2)} \cdot \hat{\mathbf{x}} & \mathbf{h}_n^{(3)} \cdot \hat{\mathbf{x}} & \mathbf{h}_n^{(4)} \cdot \hat{\mathbf{x}} \\ \mathbf{h}_n^{(1)} \cdot \hat{\mathbf{y}} & \mathbf{h}_n^{(2)} \cdot \hat{\mathbf{y}} & \mathbf{h}_n^{(3)} \cdot \hat{\mathbf{y}} & \mathbf{h}_n^{(4)} \cdot \hat{\mathbf{y}} \\ \mathbf{e}_n^{(1)} \cdot \hat{\mathbf{x}} & \mathbf{e}_n^{(2)} \cdot \hat{\mathbf{x}} & \mathbf{e}_n^{(3)} \cdot \hat{\mathbf{x}} & \mathbf{e}_n^{(4)} \cdot \hat{\mathbf{x}} \end{pmatrix}, \tag{6.2.15}$$

where $\mathbf{e}_i(n)$ and $\mathbf{h}_i(n)$ are the $i$th eigenmodes of the electric and magnetic fields respectively. The transfer matrix can be decomposed into two components,

$$\mathbf{t}(n+1) = \mathbf{M}(n+1)\,\Phi(n+1), \tag{6.2.16}$$

where $\Phi(n+1)$ is a diagonal matrix consisting of phase propagation terms that evolves the field between the two boundaries.

Here we have adopted $4 \times 4$ transfer matrices rather than the standard $2 \times 2$ transfer matrices. For electromagnetic waves travelling inside isotropic media, it is always possible to express the field as a combination of two independent modes, e.g. the $s$ and $p$ modes, and analyse the propagations of the two modes separately; the polarizations are conserved as the field passes across the interface. However, this does not hold in general for anisotropic media, where the polarisations can mix. We have begun by adopting a more general formulation, appropriate to considering wave transmission and reflection in this context.

To calculate the value of (6.2.7) at a fixed point in the medium, $z_l$ (that is, within the $l$th slice) we require expressions for both $\mathbf{R}_R$, and $\mathbf{R}_L$. These can be calculated as before in terms of the total transfer matrices associated respectively with propagation through the medium to the light and to the left of $z_l$:

$$\boldsymbol{T}_L = \prod_{n=1}^{l} \boldsymbol{t}(n), \quad \boldsymbol{T}_R = \prod_{n=l+1}^{N+1} \boldsymbol{t}(n). \tag{6.2.17}$$

These transfer matrices determine the relative magnitudes of the field components in each slice, and therefore fix the values of the reflection coefficients. *Prima facie*, the continuum case is recovered in the limit as $\delta z \to 0$ and $N \to \infty$. But here's the rub: it is precisely in this limit that the stress has been found to diverge [4, 5], and this is the crux of the conundrum we need to address. For the moment, let us proceed nonetheless by determining the relevant transfer matrices.

#### 6.2.2.3 The Boundary Conditions

The system we are considering consists of a series of homogeneous slices. The interface between two slices occurs in the $xy$–plane. Without loss of generality, we can rotate the $x$ and $y$ axes of our coordinate system so that the plane of incidence is the $xz$ plane. Consequently the wave vectors have zero $y$ components:

$$\mathbf{k}_1 = (k_{1x}, 0, k_{1z}), \quad \mathbf{k}_2 = (k_{2x}, 0, k_{2z}), \tag{6.2.18}$$

where $\mathbf{k}_1$ is the wave vector of the incident light, and $\mathbf{k}_2$ is the wave vector of the transmitted light. The frequencies of the reflected and transmitted waves must be the same as that of the incident wave, because the above conditions hold at the boundary at all times—this is only possible if the waves on either side are oscillating at the same frequency. Additionally, the conditions hold at all points on the boundary plane $z = 0$, so the changing phases of the waves on either side must agree as one moves along the boundary, i.e.

$$\mathbf{k}_1 \cdot \mathbf{r}|_{z=0} = \mathbf{k}_2 \cdot \mathbf{r}|_{z=0} \,. \tag{6.2.19}$$

As $\mathbf{r} = (x, y, z)$, and the above equations hold for all values of $x$ and $y$.

$$k_{1x} = k_{2x}. \tag{6.2.20}$$

The first component of the transmitted wave vector is therefore determined by the first component of the incident wave vector:

$$\mathbf{k}_2 = (k_{1x}, 0, k_{2z}). \tag{6.2.21}$$

The $k_z$ component of the wave vector, in subsequent slices, is determined by the wave equation.

#### 6.2.2.4 The Wave Equation

The electric field in a homogeneous slice of the system is of the form

$$\mathbf{E} = (E_x\hat{\mathbf{x}} + E_y\hat{\mathbf{y}} + E_z\hat{\mathbf{z}})e^{i(\omega t - \mathbf{k}\cdot\mathbf{r})}. \tag{6.2.22}$$

First, we derive a wave equation of general applicability. Applying Maxwell's equations

$$\nabla \times \mathbf{E} = -\frac{\partial \mathbf{B}}{\partial t}, \quad \nabla \times \mathbf{H} = \frac{\partial \mathbf{D}}{\partial t}, \tag{6.2.23}$$

to Eq. (6.2.22), we arrive at the following form of the wave equation:

$$(\epsilon)^{-1}\left\{\mathbf{k} \times \left[(\mu)^{-1} \cdot (\mathbf{k} \times \mathbf{E})\right]\right\} + \omega^2\mathbf{E} = 0. \tag{6.2.24}$$

This can be rewritten in matrix form. For a portion of a $C$-slice, which has optical properties characterised by (6.2.9), we obtain

$$\begin{pmatrix} -\frac{k_y^2}{m} - k_z^2 m + \frac{\omega^2}{m} & \frac{k_x k_y}{m} & k_x k_z m \\ \frac{k_x k_y}{m} & -\frac{k_x^2}{m} - k_z^2 m + \frac{\omega^2}{m} & k_y k_z m \\ k_x k_z m & k_y k_z m & -k_x^2 m - k_y^2 m + m\omega^2 \end{pmatrix} \begin{pmatrix} E_x \\ E_y \\ E_z \end{pmatrix} = \mathbf{0}. \tag{6.2.25}$$

#### 6.2.2.5 Eigenmodes of the $C$-Slice

For non-trivial solutions of the wave equation, we require that the determinant of (6.2.25) should be equal to zero. From the secular equation we obtain the dispersion relations of the eigenmodes. For a space with the properties of the $C$-slice, they are simple and degenerate:

$$\omega_n^{(i)} = \pm\sqrt{k_{nx}^2 + k_{nz}^2 m_n}, \tag{6.2.26}$$

for each polarisation $i$ and each slice $n$. As in the previous calculation, we may ignore dispersion by setting $m(z,\omega) = m(z)$. The eigenmodes of the electric field (prior to normalisation) are determined to be s

$$\mathbf{e}_n^{(1)} = \mathbf{e}_n^{(2)} = \left(-\frac{k_{nz} m_n^2}{k_{nx}}, 0, 1\right), \tag{6.2.27a}$$

$$\mathbf{e}_n^{(3)} = \mathbf{e}_n^{(4)} = (0, 1, 0). \tag{6.2.27b}$$

The corresponding magnetic field modes can be derived directly from the eigenmodes of the electric field via

$$\mathbf{h}_n^{(i)} = \frac{1}{\omega_n} (\mu_n)^{-1} \left(\mathbf{k}_n \times \mathbf{e}_n^{(i)}\right), \tag{6.2.28}$$

from which we obtain

$$\mathbf{h}_n^{(1)} = -\mathbf{h}_n^{(2)} = \left(0, -\frac{m_n}{\omega_n}\left\{k_{nx} + \frac{k_{nz}^2 m_n^2}{k_{nx}}\right\}, 0\right), \tag{6.2.29a}$$

$$\mathbf{h}_n^{(3)} = -\mathbf{h}_n^{(4)} = \left(-\frac{k_{nz} m_n}{\omega_n}, 0, \frac{k_{nx}}{\omega_n m_n}\right). \tag{6.2.29b}$$

#### 6.2.2.6 Reflection Coefficients in a $C$-Slice

For an interface consisting of two $C$-slice half-spaces, $\alpha$ and $\beta$, we can relate the field inside the first layer, and to the immediate left of the interface, with the field inside the second layer, and immediately right of the interface, by the zero-phase transfer matrix (6.2.14). For ease of interpretation, let us write the relationship as

$$\begin{pmatrix} E_I^t \\ 0 \\ E_{II}^t \\ 0 \end{pmatrix} = \mathbf{M} \begin{pmatrix} E_I^i \\ E_I^r \\ E_{II}^i \\ E_{II}^r \end{pmatrix}. \tag{6.2.30}$$

$E_I^i$, $E_{II}^r$, in this case, represent the waves of an arbitrarily chosen polarisation incident upon and reflected by the boundary, and $E_I^t$ the transmitted mode. For the second orthogonal polarisation, we substitute $II$ for $I$. The reflection coefficients are then:

$$r_{I\to I} = \left.\frac{E_I^r}{E_I^i}\right|_{E_{II}^i=0} = \frac{M_{24}M_{41} - M_{21}M_{44}}{M_{22}M_{44} - M_{24}M_{42}}, \tag{6.2.31a}$$

$$r_{I\to II} = \left.\frac{E_{II}^r}{E_I^i}\right|_{E_{II}^i=0} = \frac{M_{21}M_{42} - M_{22}M_{41}}{M_{22}M_{44} - M_{24}M_{42}}, \tag{6.2.31b}$$

$$r_{II\to I} = \left.\frac{E_I^r}{E_{II}^i}\right|_{E_I^i=0} = \frac{M_{24}M_{43} - M_{23}M_{44}}{M_{24}M_{42} - M_{22}M_{44}}, \tag{6.2.31c}$$

$$r_{II\to II} = \left.\frac{E_{II}^r}{E_{II}^i}\right|_{E_I^i=0} = \frac{M_{22}M_{43} - M_{23}M_{42}}{M_{24}M_{42} - M_{22}M_{44}}. \tag{6.2.31d}$$

The double-indexing acknowledges the possibility of conversion between the polarisations: for example, a wave of type I may be reflected in its orthogonal form II. For the $C$-slice, inserting expressions (6.2.27a, 6.2.27b), and (6.2.28) into matrix (6.2.15), and determining the inverse, we find the transfer matrix $\mathbf{M} = \mathbf{D}(\beta)^{-1}\mathbf{D}(\alpha)$ takes the form:

$$\mathbf{M} = \frac{1}{2}\begin{pmatrix} \frac{m_\alpha}{m_\beta^2}\left(\frac{k_{\alpha z}}{k_{\beta z}}m_\alpha + \frac{\omega_\alpha}{\omega_\beta}m_\beta\right) & \frac{m_\alpha}{m_\beta^2}\left(\frac{k_{\alpha z}}{k_{\beta z}}m_\alpha - \frac{\omega_\alpha}{\omega_\beta}m_\beta\right) & 0 & 0 \\ \frac{m_\alpha}{m_\beta^2}\left(\frac{k_{\alpha z}}{k_{\beta z}}m_\alpha - \frac{\omega_\alpha}{\omega_\beta}m_\beta\right) & \frac{m_\alpha}{m_\beta^2}\left(\frac{k_{\alpha z}}{k_{\beta z}}m_\alpha + \frac{\omega_\alpha}{\omega_\beta}m_\beta\right) & 0 & 0 \\ 0 & 0 & 1 + \frac{k_{\alpha z}\omega_\beta}{k_{\beta z}\omega_\alpha}\frac{m_\alpha}{m_\beta} & 1 - \frac{k_{\alpha z}\omega_\beta}{k_{2z}\omega_\alpha}\frac{m_\alpha}{m_\beta} \\ 0 & 0 & 1 - \frac{k_{\alpha z}\omega_\beta}{k_{\beta z}\omega_\alpha}\frac{m_\alpha}{m_\beta} & 1 + \frac{k_{\alpha z}\omega_\beta}{k_{2z}\omega_\alpha}\frac{m_\alpha}{m_\beta} \end{pmatrix}. \tag{6.2.32}$$

Evidently there is no mixing of the two polarisations in the $C$-slice. Further simplification of this matrix is possible, since $\omega_\alpha = \omega_\beta$ and

$$\sqrt{k_{\alpha x}^2 + k_{\alpha z}^2 m_\alpha^2} = \sqrt{k_{\alpha x}^2 + k_{\beta z}^2 m_\beta^2}$$
$$\Longrightarrow k_{\alpha z}^2 m_1^2 = k_{\beta z}^2 m_\beta^2.$$

We know that these factors share the same signs. Therefore we infer that

$$k_{\alpha z} m_\alpha = k_{\beta z} m_\beta. \tag{6.2.33}$$

This, of course, is to be expected: the $k_{nz}$ wave numbers in a homogeneous slice $n$ are equal to the vacuum wave number $k_{0z}$ compressed by a factor of $m_n$.

$$k_{nz} = \frac{k_{0z}}{m_n}. \tag{6.2.34}$$

With these simplications, it is clear that the matrix is diagonal, and reduces to the form

$$\mathbf{M} = \begin{pmatrix} \frac{m_\alpha}{m_\beta} & 0 & 0 & 0 \\ 0 & \frac{m_\alpha}{m_\beta} & 0 & 0 \\ 0 & 0 & 1 & 0 \\ 0 & 0 & 0 & 1 \end{pmatrix}. \tag{6.2.35}$$

Thus we find that, for all angles of incidence, the reflection coefficients (6.2.31a–6.2.31d) are identically zero; there is no internal scattering within a $C$-slice.

#### 6.2.2.7 The Force on the Plate

If we consider now the whole system, consisting of a series of $N+1$ homogeneous slices between and contiguous with the two mirrors, and the reflection coefficients associated with sending plane waves to the right and to the left of any given point $z_l$ in the cavity, it is clear that they are of the same form, but with the matrix $\mathbf{M}$ substituted by the transfer matrix $\mathbf{T_L}$ or $\mathbf{T_R}$, when calculating the left or right reflection coefficients respectively. Between the plates, these transfer matrices are diagonal matrices consisting of phase terms, being made up of products of diagonal $\mathbf{M}$ and $\mathbf{\Phi}$ matrices. For $n \in [2, N]$, the $\mathbf{M}$ matrices do not modify the reflection coefficients, and can be replaced with identity matrices, hence the contribution to the transfer matrices for $n \in (2, N)$ consists entirely of phase terms that divide into two sets: the phase terms for which $m_n = 1$ (i.e. slices of vacuum), and the phase terms for which $m_n \neq 1$ (i.e. $C$-slice material). The reflection matrices (6.2.8) are diagonalised.

We recover Casimir's original result by setting $m_n = 1\, \forall n$. This problem has already been solved, and we will not repeat the details of that calculation again. To determine the force on the left or right plate in our example, where a $C$-slice has been introduced into the cavity, we need only consider the way in which the $C$-slice modifies the relevant transfer matrices that determine the reflection coefficients. From the discussion above, it is evident without further calculation that this difference amounts to nothing more than a modification of the accumulated (imaginary) phase: slices containing $C$-slice material produce an amount of phase in their corresponding transfer matrices that differs by a factor of $m_n$ from slices of vacuum,[5] modifying the $e^{2dw}$ term in the stress to $e^{2d'w}$. The reflection coefficients at the mirrors remain unmodified during the motion. The addition of the wafer, then, simply modifies the effective length of the cavity, seen by a given frequency $\omega$, from a distance $d$ to

$$d' = d + \Delta(C_S^{-1} - 1), \tag{6.2.36}$$

[5] The $m$ factors that appear in the $M$ matrix disappear in any product of the transfer matrices spanning two sides of the device, which is situated in vacuum, where $m = 1$.

and the modified Casimir pressure can be stated exactly by simply substituting the distance paramater $d$, in the original expression for the Casimir force (1.1.23), with the effective length of the cavity $d'$. For a non-dispersive $C$-slice, the Casimir force is easily restated:

$$P = -\frac{\hbar c \pi^2}{240 d'^4} = -\frac{\hbar c \pi^2}{240\left(d + \Delta(C_S^{-1} - 1)\right)^4}. \quad (6.2.37)$$

## 6.3 Applications of the $C$-slice

Quantum stiction, due to the 'stickiness' of the Casimir effect, has been acknowledged as a serious engineering problem for micro and nano–machinery [3]. At such length scales, Casimir-Lifshitz forces are no longer negligible and can lead to unwanted attraction between, and adhesion of, material parts in the device. Although immersion in a suitably dense liquid may reduce the phenomenon of quantum stiction (by effectively increasing the optical distance between the parts) this would introduce viscous forces into the system, among other complications. For micromachines that cannot be thus immersed (for example, where the parts need to move in relation to each other) the freedom of a surrounding vacuum remains necessary. An idealised $C$-slice seems to suggest itself as one possible solution to this problem. Without modifying the physical dimensions of the cavity, or introducing new surfaces for interaction, the *effective* size of the cavity can be made arbitrarily large, and the Casimir force made arbitrarily small, by interposing a thin layer or wafer made to the appropriate specifications (6.2.1).[6] A similar proposal for tackling stiction was made in [13], using a negatively refracting medium to produce a repulsive Casimir force, with the disadvantage that optical pumping would be required. The $C$-slice does not require optical pumping. However, as we observed earlier, the quantum Casimir effect is a broadband phenomenon, and it seems unlikely that the necessary condition of impedance-matching could be secured for a sufficiently large portion of the electromagnetic spectrum, though an argument to the contrary can be found in [13]. Whether or not a material could be made that approximated the effects of the $C$-slice is a problem we will not examine here. For the thermal Casimir effect, however, it has been observed that, as the temperature increases, contributions to the force are increasingly distributed around a characteristic frequency [20]; it is perhaps more conceivable that an impedance-matched metamaterial implementing the $C$-slice could be designed for tuning the thermal Casimir effect. However, there are applications outside of the domain of Casimir Physics. By virtue of its non-reflective properties and its capacity to change the measure of optical distance, the compressive transformation implemented by a $C$-slice can be used to reduce the profile of a lens [21], for example. Perhaps there may also be some use for the $C$-slice in a

---

[6] In this case, a homogeneous $C$-slice would be sufficient to achieve the desired effect, and should be engineered so that $m < 1$.

**Fig. 6.2** The Casimir force produces 'quantum stiction' in micro and nanomachinery, causing tiny machine parts to stick together. *Image credit* http://mems.sandia.gov

laser system, should there be occasion to modify the resonant frequency of a cavity without disturbing the cavity walls. We must leave the development of such ideas for discussion elsewhere, however.

## 6.4 Summary Remarks

In this chapter, we have imagined introducing an inhomogeneous, anisotropic transformation medium into a cavity that implements a simple distortion of the laboratory coordinate system, effectively compressing or expanding space along one axis. For an idealised $C$-slice, configured to expand the space between the mirrors, the Casimir force can in theory be reduced arbitrarily (Fig. 6.2).

Initially, it seemed this *Gedankenexperiment* could contradict our earlier conclusions, in Chap. 5: it appears the Casimir stress in an inhomogeneous medium is cutoff-dependent. However, the inhomogeneous $C$-slice merely modifies the optical distance between the walls of an empty cavity. In an empty cavity, the Casimir stress is independent of the cut-off.

We have established fairly straightforwardly that there is no contradiction here. The Casimir stress in an inhomogeneous medium may depend on details which, as yet, remain unincorporated within Casimir theory. However, in such cases there is scattering within the medium; the stress diverges [4], and the force is cutoff-dependent [5], because the reflection coefficients fail to fall off fast enough in the limit of high wave numbers, as discussed in Sect. 5.4.1. However, in the case of the $C$–slice, the reflection coefficients are precisely zero for all angles of incidence, and the stress does not diverge. Such 'transformation media' merely change the measure of optical distance without introducing additional scattering events. Therefore, it is possible to determine the Casimir forces for systems incorporating transformation media, even when they involve continuously changing dielectric properties. In this chapter, we have essentially identified a subset of idealised inhomogeneous systems for which Lifshitz theory may still make meaningful predictions.[7] In the example we have considered, where we interpose a $C$-slice between two parallel plates, or expand it to fill the entire volume of the cavity, the medium changes the Casimir force by modifying the effective size of the cavity. However, the force on the cavity's walls remains attractive and finite.

## References

1. H.B.G Casimir, Koninkl. Ned. Akad. Wetenschap. **51**, 793 (1948)
2. E.M. Lifshitz, Zh. Eksp. Teor. Fiz. **29**, 94 (1955)
3. F. Capasso, J.N. Munday, *Casimir Physics: Lecture Notes in Physics*, vol. 834 (Springer, Berlin, 2011)
4. W.M.R. Simpson, S.A.R. Horsley, U. Leonhardt, Phys. Rev. A, **87**, 043806 (2013)
5. S.A.R. Horsley, W.M.R. Simpson, Phys. Rev. A **88**, 013833, (2013)
6. S.A.R. Horsley, T.G. Philbin, New J. Phys. **16**(1), 013030 (2014)
7. W.M.R. Simpson, Phys. Rev. A **88**, 063852 (2013)
8. U. Leonhardt, T.G. Philbin, *Geometry and Light: The Science of Invisibility* (Dover, New York, 2010)
9. J.B. Pendry, D. Schuring, D.R. Smith, Science **312**, 1780 (2006)
10. T.H. Boyer, Phys. Rev. A **9**, 2078 (1974)
11. K.A. Milton, Ann. Phys. (N.Y.) **127**, 49 (1980)
12. N.B. Kundtz, R. Smith, J.B. Pendry, Proc. IEEE **99**, 1622–1633 (2011)
13. U. Leonhardt, T.G. Philbin, New J. Phys. **9**, 254 (2007)
14. F.S.S. Rosa, D.A.R. Dalvit, P.W. Milonni, Phys. Rev. Lett. **100**, 183602 (2008)
15. C. Genet, A. Lambrecht, S. Reynaud, Phys. Rev. A **67**, 043811 (2003)
16. M. Born, E. Wolf, *Principles of Optics*, 7th ed. (Cambridge University Press, Cambridge, 1999)
17. I.A. Shelykh, V.K. Ivanov, Int. J. Theor. Phys. **43**, 477 (2004)
18. M. Artoni, G.C. La Rocca, F. Bassani, Phys. Rev. E **72**, 046604 (2005)
19. J. Hao, L. Zhou, Phys. Rev. B **77**(9), 094201 (2008)
20. M. Li, R.-X. Miao, Y. Pang, Opt. Express **18**(9), 9026–9033 (2010)
21. D.A. Roberts, N. Kundtz, D.R. Smith, Opt. Express **17**(19), 16535-16542 (2009)
22. U. Leonhardt, W.M.R. Simpson, Phys. Rev. D **84**, 081701 (2011)

[7] The only other example where a finite Casimir force has been calculated for an inhomogeneous medium that the author is aware of is [22]. In this case, however, the inhomogeneous system did not constitute a transformation medium.

# Chapter 7
# The Casimir Force in Maxwell's Fish-Eye

*Let no one enter who does not know geometry.*
Inscription over the door of Plato's Academy

## 7.1 The Paradox of Maxwell's Fish-Eye

In this chapter we will consider a conundrum created by the case of the Casimir force in Maxwell's fish-eye.[1] And this time we cannot offer a complete resolution. Our last problem is a perplexing one. Maxwell's fish-eye is an inhomogeneous metamaterial with some remarkable properties. The notion of materials distorting the geometry of space for light is one we have touched on already. In the case of Maxwell's fish-eye, this metamaterial induces a stereographic projection of the hypersphere onto $3d$ flatspace, and in so doing threatens to warp the classic procedure of regularisation in Casimir Physics. Once again, as we will see, adopting a geometric viewpoint leads to a paradox: the symmetry properties of the fish-eye (without a mirror) exclude the possibility of a Casimir force; however, the stress tensor is infinite throughout the medium, and likewise the predicted self-force. 'These things ought not so to be'. In grappling with this problem, we find that we can in fact uncover a solution for the Casimir stress in an inhomogeneous system containing the medium of a fish-eye, by exploiting its geometric properties, but it is an answer that leads to more questions.

[1] The main results in this chapter were published in [1].



W.M.R. Simpson, *Surprises in Theoretical Casimir Physics*, Springer Theses,
DOI 10.1007/978-3-319-09315-4_7

## 7.2 The Geometry of the Fish-Eye

The fish-eye is an impedance matched[2] inhomogeneous medium with an electric permittivity and magnetic permeability equal to

$$\varepsilon = \mu = n = \frac{2n_1}{1 + (r/a)^2}. \tag{7.2.1}$$

For simplicity, we will set $n_1 = 1,\ a = 1$, throughout this chapter, corresponding to the case of a fish-eye of unit radius with a background refractive index of unity:

$$n = \frac{2}{1 + r^2}. \tag{7.2.2}$$

### 7.2.1 The Stereographic Projection

From Fermat's principle, we know that light rays travel extremal optical paths with path lengths $s$ measured by the refractive index $n$,

$$s = nl. \tag{7.2.3}$$

In other words, light travels along geodesics. We can think of optical materials as modifying the geometry of space for light. Significantly, the transformation induced by the fish-eye does not constitute merely a reconfiguration of flat coordinate space, unlike the $C-$slice. The geometry of the fish-eye has an intrinsic curvature, as can be verified from the computation of the curvature scalar [2], which is given by

$$R = \frac{8(\partial_z n)(\partial_{z^*} n)}{n^4} - \frac{8\partial_z \partial_{z^*} n}{n^3}, \quad n = \frac{2}{1 + |z|^2} \quad \Rightarrow \quad R = 2 \neq 0. \tag{7.2.4}$$

But what are the geodesics of this geometry? To discover this, we must consider just how the fish-eye modifies the measure of space, effectively absorbing the material of the fish-eye within the geometric properties of an empty 'virtual space', described by a primed coordinate system in which $n' = 1$. For convenience, let us adopt a complex number representation of the coordinate system of physical space, $\zeta = x + iy$, and restrict ourselves to considering the curved space induced by a 2d fish-eye, for which we require 3 dimensions in Cartesian space $(x', y', z')$. The line element is defined

$$\mathrm{d}s'^2 = \mathrm{d}x'^2 + \mathrm{d}y'^2 + \mathrm{d}z'^2. \tag{7.2.5}$$

[2] Impedance matching is a condition imposed in [2] for securing a virtual geometry for electromagnetic fields.

We have already hinted that the relationship between the two coordinate systems is given by a stereographic projection. Adopting the ansatz

$$\zeta = \frac{x' + iy'}{1 - z'}, \tag{7.2.6}$$

which is more conveniently expressed in its inverse form,

$$x' + iy' = \frac{2\zeta}{1 + |\zeta|^2}, \quad z' = \frac{|\zeta|^2 - 1}{|\zeta|^2 + 1}, \tag{7.2.7}$$

we determine that

$$\mathrm{d}\left(x' + iy'\right) = \mathrm{d}\left(\frac{2\zeta}{1 + |\zeta|^2}\right) = \frac{2(\mathrm{d}\zeta - \zeta^2 \mathrm{d}\zeta^*)}{(1 + |\zeta|^2)^2}, \tag{7.2.8a}$$

$$\mathrm{d}z' = \mathrm{d}\left(\frac{|\zeta|^2 - 1}{|\zeta|^2 + 1}\right) = \frac{2(\zeta^* \mathrm{d}\zeta + \zeta \mathrm{d}\zeta^*)}{(1 + |\zeta|^2)^2}, \tag{7.2.8b}$$

where the differential operator has been computed using the substitution

$$\mathrm{d} \to \mathrm{d}\zeta\, \partial_\zeta + \mathrm{d}\zeta^*\, \partial_{\zeta^*}. \tag{7.2.9}$$

The line element (7.2.5) can be conveniently rewritten in the form

$$\mathrm{d}s'^2 = \mathrm{d}\left(x' + iy'\right) \mathrm{d}\left(x' - iy'\right) + \mathrm{d}z'^2. \tag{7.2.10}$$

Inserting (7.2.8a, 7.2.8b) into (7.2.10) produces

$$\mathrm{d}s'^2 = \frac{4}{(1 + |\zeta|^2)^4} \left\{(\mathrm{d}\zeta - \zeta^2 \mathrm{d}\zeta^*)(\mathrm{d}\zeta^* - \zeta^{*2} \mathrm{d}\zeta) + (\zeta^* \mathrm{d}\zeta + \zeta \mathrm{d}\zeta^*)^2\right\}. \tag{7.2.11}$$

Simplifying, we obtain

$$\mathrm{d}s'^2 = \left(\frac{2}{1 + |\zeta|^2}\right)^2 \mathrm{d}\zeta \mathrm{d}\zeta^*. \tag{7.2.12}$$

Making the substitutions $|\zeta|^2 \to r^2$ and $\mathrm{d}\zeta \mathrm{d}\zeta^* \to \mathrm{d}l^2 = \mathrm{d}x^2 + \mathrm{d}y^2$, we find that

$$\mathrm{d}s^2 = \mathrm{d}s'^2 = \left(\frac{2}{1 + r^2}\right)^2 \mathrm{d}l^2 = n^2 \mathrm{d}l^2. \tag{7.2.13}$$

Thus a medium with the properties of Maxwell's fish-eye in 2d performs the stereographic projection of the surface of a sphere to the 2d plane (Fig. 7.1). Generalising to 3d, the fish-eye induces the geometry of the surface of the hyper-sphere. In the curved space of the fish-eye, the 'shortest paths' are the great circles (see Fig. 7.2),

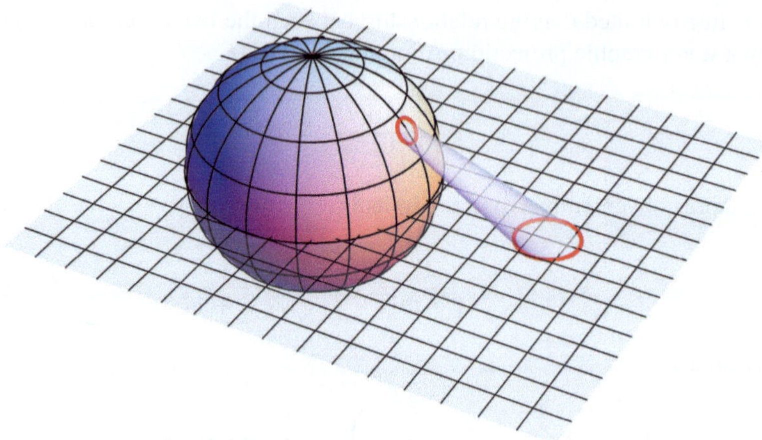

**Fig. 7.1** In 2d, the fish-eye induces the stereographic projection of the surface of a sphere onto a plane. *Image credit* Geometry and Light [2]

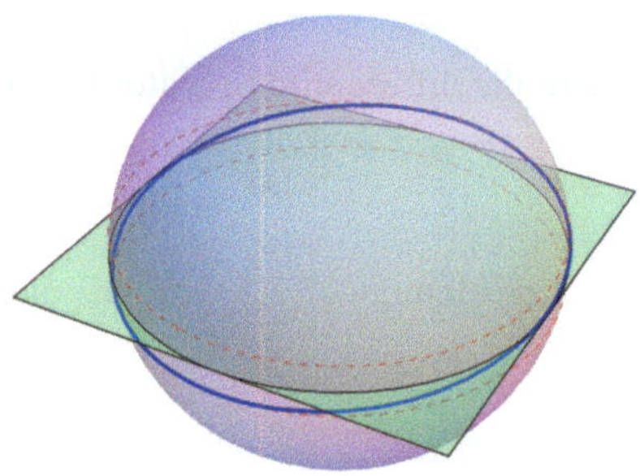

**Fig. 7.2** In 2d, light travels in the virtual space of the fish-eye along the great circles of the surface of the sphere

and light rays emitted from an arbitrary point $r_0$ will travel along one of the great circles before passing through the antipodal point $r_0' = -r_0/r_0^2$. In fact, circles on the sphere (or hypersphere) map to circles on the plane (or block) in the stereographic projection.[3] It follows that light goes around in circles in Maxwell's fish-eye.

### 7.2.2 The Force in the Fish-Eye

It is clear from the symmetries of the case that there is no net Casimir force in Maxwell's fish-eye; such a force could have no preferred direction.[4] However, the fish-eye constitutes an inhomogeneous medium, and we have amassed some

[3] See Appendix and [2].

[4] If it is suggested that a force might conceivably act radially outwards, in all directions, we simply remind the reader that the force in physical space must be the same as the force in the virtual free-space, and in virtual space we are dealing with a surface.

evidence in Chaps. 4 and 5 suggesting that the theory of the Casimir effect suffers from a systemic divergence problem in such cases. For the case of non-reflective transformation media, discussed in Chap. 6, the divergence does not bite. However, Maxwell's fish-eye implements a geometry with non-zero curvature, and a cavity filled with fish-eye material will involve non-zero reflection coefficients, however it is 'sliced'.[5]

Calculations (which we will not repeat here) have confirmed that Lifshitz theory predicts an infinite force for the fish-eye.[6] Still, it may be possible to extract a meaningful result from a Casimir problem involving the fish-eye by applying some of our knowledge of the fish-eye's geometric properties. The precise form that our thought-experiment will take is partly motivated by a problem that Casimir posed himself.

## 7.3 Maxwell's Fish-Eye and the Finestructure Constant

### *7.3.1 Casimir's Semi-classical Model of the Electron*

In 1953 Casimir suggested an intriguing model that could explain the stability of charged particles and the value of the finestructure constant [5]. The argument goes as follows. Imagine a particle, like the electron, as an electrically charged hollow sphere. Two forces are acting upon it: the Coulomb force (producing electrostatic repulsion) and the force of the quantum vacuum (presumed to be attractive). The stress $\sigma$ of the quantum vacuum on a spherical shell of radius $a$ must be given by a dimensionless constant times $\hbar c/a^4$ on purely dimensional grounds—the quantum stress is an energy density proportional to $\hbar$, and $\hbar c/a^4$ has the units of an energy density. Now, the electrostatic energy of the sphere is proportional to the square $e^2$ of its charge and is also inversely proportional to $a^4$ [6]. It follows that the Casimir force should balance the electrostatic repulsion, regardless of how small the distance parameter $a$ may be, as long as $e^2/(\hbar c)$ assumes a certain value given by the strength of the Casimir force. This strength depends on the internal structure of the particle—the fact that it is a spherical shell—but not on its size, which could be imperceptibly small. Thus Casimir's model, albeit crude, suggests a seductively simply explanation for both the finestructure constant $e^2/(\hbar c)$ and the stability of charged elementary particles.

Unfortunately Casimir's hopes for connecting the quantum vacuum to the stabilising Poincare stresses required in the classical electron model proved to be short-lived. In a seminal paper by Boyer in 1968 it was argued at length that the sign of the zero-point energy is, in fact, the opposite to that proposed in Casimir's model, and consequently the Casimir force on a spherical shell is repulsive, not attractive; the

[5] See the discussion in Sect. 6.4

[6] The modified regulariser proposed in [3] has also been tested, with negative results [4].

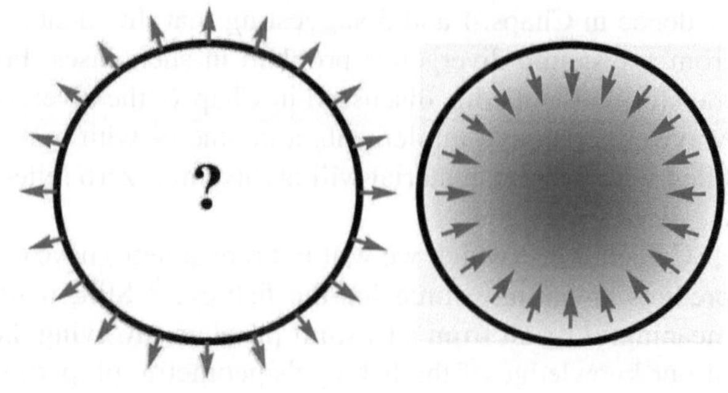


**Fig. 7.3** The Casimir force on a spherical shell is repulsive—or is it? We assumed the shell to be filled with a medium (*right*) and found an attractive Casimir stress in the material. The *shades of grey* indicate the profile of the medium

force of the vacuum works with the Coulomb force to *expand* the sphere, rather than stabilise it [7]. In Boyer's own words, this seemed 'a most melancholy' conclusion, subsequently corroborated by other attempts at the same calculation [8]. Indeed, the spherical shell has become an archetype for a shape that causes Casimir repulsion (Fig. 7.3).

However, doubts have been lingering over the question of whether the repulsive force of the shell may be an artifact of the simple model used, for the following reason: the bare stress of the quantum vacuum is always infinite and this infinity is removed by regularisation procedures. The most plausible procedure involves considering the relative stress between (or inside) macroscopic bodies. Yet an infinitely thin sphere does not represent an extended macroscopic body, nor multiple bodies. Suppose the physically relevant vacuum stress of an extended spherical shell tends to infinity in the limit when the shell becomes infinitely thin and infinitely conducting. In this case the renormalization would remove this physically meaningful infinity, producing a finite result that may very well have the wrong sign.[7]

### *7.3.2 The Fish-Eye with a Mirror*

Suppose we consider a modification of Casimir's model. Imagine the spherical shell—still infinitely conducting and infinitely thin—is no longer hollow, but filled with an impedance matched medium of gradually varying electric permittivity $\epsilon$ and magnetic permeability $\mu$, with the refractive index profile of Maxwell's fish-eye (7.2.1). For the boundary, we introduce a perfect mirror at $r = 1$. In 2d, this amounts to placing a mirror on the equator of the sphere in virtual. In 3d, the mirror is a 2d surface situated on the 4d hypersphere.

In this way we have extended the shell to a macroscopic body where the Casimir stress (2.135) gradually builds up, whilst keeping the idealised boundary intact. As before, this model possesses spherical symmetry, since we assume that $\epsilon$ and $\mu$

[7] The authors of [9], for example, claim 'Lifshitz theory shows that the self-force is in fact inwardly directed and infinite'.

depend only on the distance $r$ from the center of the sphere. We continue to expect no Casimir force in the center, therefore, because in a spherically symmetric medium the center does not distinguish any direction for a force vector to point to. Our object now is to calculate the Casimir stress tensor $\sigma(r)$ which will be radially symmetric and may change with increasing $r$. Lifshitz' result (2.164), which is written in terms of reflection coefficients and has served us well up till this moment, is no longer applicable; we must turn instead to the more general formulation of the Casimir stress, written in terms of the full electromagnetic Green function (2.135).

#### 7.3.2.1 The Green Function of the Fish-Eye

The Green function of the bare fish-eye satisfying Maxwell's equations and determining the behaviour of electromagnetic waves in the medium has been derived in [10]. The derivation is rather lengthy and involved and we will not recite it here, but it is documented fully and accessibly in [2]. The Green function may be succinctly expressed in the form

$$G(r, r') = \frac{\nabla \times n(r_m)\nabla \otimes \nabla' D(r_m) \times \overleftarrow{\nabla'}}{n(r)n(r')k^2}, \tag{7.3.1}$$

where $D$ describes a scalar wave on the hypersphere,

$$D(r_m) = \frac{1}{8\pi}\left(r_m + \frac{1}{r_m}\right)\frac{\sinh(2ik \operatorname{arccot}[r_m])}{\sinh(\pi ik)}, \tag{7.3.2}$$

and is a function of the Möbius-transformed radius, defined by

$$r_m = \frac{|\mathrm{r} - \mathrm{r}'|}{\sqrt{1 + 2\mathrm{r} \cdot \mathrm{r}' + r^2 r'^2}}. \tag{7.3.3}$$

Any rotation on the hypersphere can be described by a Möbius transformation in physical space. It can be shown that this form of the tensor is Möbius invariant. It follows that (7.3.1) holds for any point in space in which light can be found in the fish-eye. If we imagine a succession of flashes from a source point in the fish-eye, the combination of these two functions, $G$ and $D$, dictates a reflection at the antipodal point that will return each wave back to its source. The Green function of the fish-eye is a bitensor that describes the electric field of a stationary wave with wave number $k$, emitted by an elementary dipole placed at $r'$ and measured at $r$; $r$ and $r'$ are distances in physical space from the centre of the fish-eye. For our purposes, we will be considering waves with imaginary frequencies $\kappa$.

#### 7.3.2.2 The Green Function of the Fish-Eye with a Mirror

We wish to calculate the Casimir force in the fish-eye with a mirror. In the geometrical picture associated with Maxwell's fish eye, the electromagnetic wave propagates on the surface of the hypersphere from a source at $r'$ to a spectator at $r$ in stereographic coordinates with distance $\arctan r_m$, and $D$ denotes the Green function of a conformally coupled scalar field [2]. The effect of the mirror is described by an adaptation of the method of images [6] on the hypersphere. There the mirror lies on the equator (a 2-dimensional surface for the 4-dimensional hypersphere). Our method is to subtract from the incident Green function $G_i$ the electromagnetic wave generated by the image source on the hypersphere $G_r$,

$$G = G_i(r) - G_r(r). \tag{7.3.4}$$

This reflected field is the mirror image of the original field. In stereographic projection, the reflection at the equator corresponds to the transformation[8]

$$\mathrm{r} \rightarrow \frac{1}{r^2}\mathrm{r}, \tag{7.3.5}$$

or, in spherical coordinates, $r \rightarrow r^{-1}$. To obtain the reflected wave $G_r$, we thus perform the coordinate transformation $r = r(r')$ with $r' = r^{-1}$, and then replace $r'$ by $r$ in the incident wave $G_i$. As the Casimir stress in a spherically symmetric medium must itself be spherically symmetric, we calculate $G_i(r^{-1})$ only in the $x$ direction, *i.e.* we put $y = z = 0$, $x' = 1/x$, and $y' = z' = 0$. The Green function (7.3.1) is rather complicated. With the aid of Mathematica, we obtain

$$G_i(r^{-1}) = \frac{(1+x^2)^2}{16x^2\kappa^2\xi}\begin{pmatrix} 2D'(\xi) & 0 & 0 \\ 0 & D'(\xi) + \xi D''(\xi) & 0 \\ 0 & 0 & D'(\xi) + \xi D''(\xi) \end{pmatrix}, \tag{7.3.6}$$

where $\xi = \frac{1}{2}\left(-2 + \frac{1}{x^2} + x^2\right)^{1/2}$. The inversion has preserved the line element of Maxwell's fish eye, and hence the transformed scalar Green function remains a solution of the wave equation. Before arriving at the reflected wave, however, we also need to transform the field components of the Green tensor $G_i$. This is effected by applying the Jacobian

[8] This is simply the inversion in the unit sphere as a mirror transformation of the spectator points. The inversion takes any point $P$ (other than the origin $O$) to its image $P'$, but also takes $P'$ back to $P$, so that the result of applying the same inversion twice is the identity transformation on all the points of the plane other than $O$. It follows that the inversion of any point inside the reference circle must lie outside it—in this case, beyond the mirror.

$$P = \frac{\partial \mathrm{r}'}{\partial \mathrm{r}} = \begin{pmatrix} \frac{\partial x'}{\partial x} & \frac{\partial x'}{\partial y} & \frac{\partial x'}{\partial z} \\ \frac{\partial y'}{\partial x} & \frac{\partial y'}{\partial y} & \frac{\partial y'}{\partial z} \\ \frac{\partial z'}{\partial x} & \frac{\partial z'}{\partial y} & \frac{\partial z'}{\partial z} \end{pmatrix}, \tag{7.3.7}$$

such that

$$G_r(r) = PG_i(r^{-1}). \tag{7.3.8}$$

Since reflection results in a phase shift of $\pi$, we subtract the reflected field from the original one. In this way we obtain the Green function for the fish-eye with a mirror:

$$G = G_i(r) - PG_i(r^{-1})\,. \tag{7.3.9}$$

For the coordinate transformation (7.3.5), the Jacobian (7.3.7) is equal to

$$\frac{1}{r^4}\begin{pmatrix} -x^2+y^2+z^2 & -2xy & -2xz \\ -2xy & x^2-y^2+z^2 & -2yz \\ -2xz & -2yz & x^2+y^2-z^2 \end{pmatrix}, \tag{7.3.10}$$

which can be succinctly written in the form

$$P = \frac{1}{r^2} * \mathbf{1}_3 - \frac{2}{r^4}\mathrm{r}\otimes\mathrm{r}. \tag{7.3.11}$$

In this case, $P$ has the simple form

$$P = \frac{1}{x^2}\begin{pmatrix} -1 & 0 & 0 \\ 0 & 1 & 0 \\ 0 & 0 & 1 \end{pmatrix}. \tag{7.3.12}$$

Hence the second term in (7.3.8) is

$$-PG_i = \frac{(1+x^2)^2}{16x^4\kappa^2\xi}\begin{pmatrix} 2D'(\xi) & 0 & 0 \\ 0 & -D'(\xi)-\xi D''(\xi) & 0 \\ 0 & 0 & -D'(\xi)-\xi D''(\xi) \end{pmatrix}. \tag{7.3.13}$$

One verifies that the transversal components of $G$ vanish at $r = 1$, as they should at a perfectly reflecting electric mirror.

#### 7.3.2.3 Deducing Field Equality

In order to calculate the stress function for this case, we will need to determine the correlation function $\tau$, defined in equation (2.136), including both its electric component $\tau_E$ (the first term) and magnetic component $\tau_M$ (the second term). The determination of $\tau_M$ for the Green function we obtained above turns out to be computationally intensive. It is possible to circumvent this calculation, however, by demonstrating *a priori* the equality of the electric and magnetic components for

1. the impedance matched incident Green function,
2. the reflected Green function, with or without impedance matching.

In fact, it is possible to show that this equality holds in general for impedance matched cases. For this, we represent the impedance matched Green function $G$ as

$$G = \mathrm{G} + \frac{\delta(r-r')\,\mathbf{1}_3}{n\kappa^2}, \tag{7.3.14}$$

and obtain from the wave equation

$$\nabla \times \frac{1}{n}\nabla \times \mathrm{G} + n\kappa^2 \mathrm{G} + \nabla \times \frac{1}{n}\nabla \times \frac{\delta(r-r')\,\mathbf{1}_3}{n\kappa^2} = 0. \tag{7.3.15}$$

Reexpressing the double curl of the delta-function term in terms of derivatives with respect to $r$ and $r'$, we obtain

$$\nabla \times \frac{1}{n}\nabla \times \mathrm{G} + n\kappa^2 \mathrm{G} = -\frac{\nabla \times \delta(r-r')\,\mathbf{1}_3 \times \overleftarrow{\nabla}'}{n(r)n(r')\kappa^2}. \tag{7.3.16}$$

Notice that the magnetic Green function, defined as

$$G_M = -\frac{\nabla \times G \times \overleftarrow{\nabla}'}{n(r)n(r')\kappa^2} \tag{7.3.17}$$

obeys the same wave Eq. (7.3.16) up to a delta-function term.[9] Consequently, $G_M$ agrees with $G$ apart from a delta-function term, but such a term does not matter in the correlation function (2.136) where we regard $r \neq r'$ before we take the limit $r' \to r$.

Concerning our second claim, consider $G_M$ defined by Eq. (7.3.17) for the transformed Green function $\mathrm{G}_r$ that describes the reflected wave and rename $r$ as $r'$ (we recall that $G_r$ is the result of a coordinate transformation). Then we make use of two geometrical facts. First, $n(r)^{-1}\nabla \times G_i$ defines a one-form with respect to the effective geometry with line element $n\,dl$ [2]. Second, this one-form is invariant under the transformation $r' = r^{-1}$, because the line element $n\,dl$ is invariant. Therefore we can read $n(r')^{-1}\nabla' \times G_r$ as the coordinate-transformed $n(r)^{-1}\nabla \times G_i$. Hence we can also read the magnetic Green function of $G_r$ (with $r$ renamed as $r'$) as the coordinate-transformed $G_M$ of $G_i$. For $G_i$ the entire medium is impedance-matched, and in such a case we have already established that the magnetic Green function agrees with the

[9] It is equivalent to a duality transform of the electric Green function, which obeys the same wave equation.

electric one up to a delta-function term. Consequently, the same must be true for the transformed Green function $G_r$ that describes the reflection at the mirror. Separating the two terms in the correlation function (2.136) into electric and magnetic components, $\tau_E$ and $\tau_M$, we conclude that $\tau_E = \tau_M$ and obtain[10]

$$\tau = -\frac{2\,\hbar c\, n}{\pi} \int_0^\infty \kappa^2 \mathrm{G}(r, r'; \mathrm{i}\kappa)\, \mathrm{d}\kappa, \tag{7.3.18}$$

where we have transferred the prefactor and the integrals from (2.135) to (2.136). We are now ready to calculate the Casimir stress $\sigma$ from the electromagnetic Green function, according to Eqs. (2.135) and (7.3.18).

#### 7.3.2.4 Regularising the Green function

The unregularised stress tensor is, of course, infinite. As we have discussed before, the standard procedure in Lifshitz theory for regularising the Casimir force is to subtract from the Green function an auxiliary Green function corresponding to an infinite homogeneous medium sharing the optical properties of the system at the point of measurement:

$$\tilde{G} = G(r) - G_0(r). \tag{7.3.19}$$

This constitutes a kind of renormalisation; the physical contribution to the Casimir stress is understood to arise due to inhomogeneities in the surrounding material, so the Casimir force is 'renormalised' to zero for a homogeneous space. Unfortunately, as we have found, this method does not remove all the divergences from an inhomogeneous medium, such as Maxwell's fish-eye. Moreover, modifications to the standard procedure for regularising the Casimir force must be physically well-motivated.[11]

However, in this case we appear to occupy a privileged position: the Green function $G$ of an infinitely extended fish-eye medium without a mirror corresponds to the Green function on the entire surface of the hypersphere, which is a uniform space. It can only produce a uniform vacuum stress $\sigma_0$ that does not contribute to the Casimir-force density $\nabla \cdot \sigma$. There is no Casimir force in the bare fish-eye. It follows that the renormalised Green function

$$G(r) - G_0(r) = G_i(r) - G_0(r) - PG_i(r^{-1}) \tag{7.3.20}$$

has a redundant component

$$G_i - G_0 \tag{7.3.21}$$

[10] This identity has been confirmed to hold for the fish-eye using Mathematica.

[11] In Sect. 4.4 we discussed an example of a modification to the Casimir force that would remove divergences in inhomogeneous media, but at the expense of modifying the value of the force itself.

that we can ignore; we only need to focus on the reflected part of the radiation field. We will thus consider only $-P\,G_i(r^{-1})$ in the total Green function (7.3.9) in our calculation of the correlation function and the stress tensor.

#### 7.3.2.5 Calculating the Correlation Function

In order to compute this *bona fide* Casimir stress, it is judicious to perform the integration over $\kappa$ before the differentiations in (7.3.13). We obtain for the scalar Green function

$$\int_0^\infty D\,\mathrm{d}\kappa = \frac{1+\xi^2}{16\pi\,\xi^2}. \tag{7.3.22}$$

It follows that

$$D' = \frac{1}{8\pi\xi} - \frac{1+\xi^2}{8\pi\xi^3}, \quad D'' = -\frac{3}{8\pi\xi^2} + \frac{3\left(1+\xi^2\right)}{8\pi\xi^4}, \tag{7.3.23}$$

and that

$$D'(\xi) - \xi D''(\xi) = \frac{1}{8\pi\xi} - \frac{1+\xi^2}{8\pi\xi^3} + \xi\left(-\frac{3}{8\pi\xi^2} + \frac{3\left(1+\xi^2\right)}{8\pi\xi^4}\right) = \frac{1}{4\pi\xi^3}. \tag{7.3.24}$$

For the correlation function $\tau$, we thus obtain

$$\tau = \frac{c\left(1+x^2\right)\hbar}{16\pi^2x^4\xi^4}\mathbf{1}_3. \tag{7.3.25}$$

All three eigenvalues of $\tau$ are identical in the $x$-direction and, by virtue of spherical symmetry, they must be identical in all directions. Equation (7.3.25) is therefore valid everywhere in the medium. For the stress function (2.135) this yields

$$\sigma = \tau - \frac{1}{2}\mathrm{Tr}\,[\tau] = -\frac{c\left(1+x^2\right)\hbar}{64\pi^2x^4\xi^4}\mathbf{1}_3. \tag{7.3.26}$$

Substituting for $\xi$, and reintroducing the refractive index profile $n$, we obtain

$$\sigma = -\frac{\hbar c}{\pi^2 n\,(1-r^2)^4}\mathbf{1}_3. \tag{7.3.27}$$

This is a remarkable result: we have found a simple, exact expression for the Casimir stress in a system composed of spherical mirrors and filled with the inhomogeneous material of Maxwell's fish-eye – the spherical analogue of a Casimir cavity filled with an inhomogeneous medium. The stress is negative and falls monotonically, so

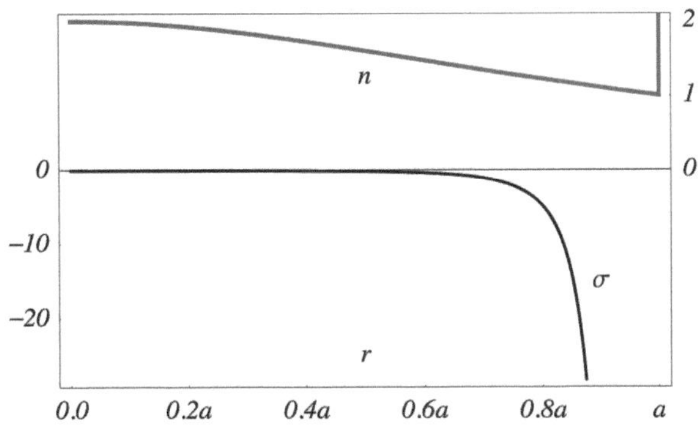


**Fig. 7.4** Index profile $n(r)$ (*grey curve*) of the medium inside the shell and the resulting vacuum stress $\sigma(r)$ (*black curve*, in units of $\hbar c/a^4$). As $r \to a$, the stress $\sigma \to \infty$

the Casimir-force density $\nabla \cdot \sigma$ is always attractive in this model. Analytic solutions of this kind are extremely rare. This result is easily generalised to the case of a system of arbitrary radius $a$:

$$\sigma = -\frac{\hbar c}{\pi^2 a^4 n\,(1 - r^2/a^2)^4}\mathbf{1}_3. \tag{7.3.28}$$

Close to the mirror $r \to a$ the stress and the force tend to infinity, however. At the mirror itself $r = a$ the stress is actually infinite; it is still not possible to predict the actual force on the mirror (see Fig. 7.4).

## 7.4 Speculations on Regularisation

What could the cause of this residual infinity be? At least three possibilities suggest themselves: it could be because we failed to incorporate spatial dispersion; perhaps it is due to the idealisation of a perfect curved mirror; or it may be the result of using the wrong regularisation. It is worth discussing each of these possibilities, though it seems only fair to caution the reader that we are teetering on the edge of what we understand, and what follows can at best be construed as informed speculation.

As touching the first possibility, it was tentatively suggested in Part 2 that the divergence in the Casimir stress that plagues inhomogeneous media may have its roots in the general failure to fully incorporate their dispersive properties—including the dependence of the dielectric response on *wavenumber* as well as frequency. On the other hand, we might appeal to the fact that, in some sense, the problem of a fish-eye enclosed in a mirror is analogous to Casimir's problem of the two parallel mirrors, in that we can reconceive the former as an empty curved space in a spherical mirror. On the surface of the hypersphere, excluding the mirror, $n'$ is homogeneous and equal to unity. Nevertheless, the sphere has curvature, which represents something of the inhomogeneity of the medium, and this feature cannot be simply removed by a coordinate transformation, as it was in Chap. 6. A residue of the inhomogeneous nature of the problem survives—amassing its effects, perhaps, around the spherical mirror in the limit as $r \to a$.

This thought brings us to consider the second possibility: our problem clearly involves a rather crude model of the mirror that encloses the medium. For instance, as in Casimir's original thought experiment, the mirror is perfectly reflective for all frequencies. In reality, mirrors only reflect imperfectly for a certain range of frequencies. On the other hand, in Casimir's calculation, as we have seen, the regularisation can be construed as implementing a crude form of dispersion, and we do in fact recover a finite result, whether the problem is calculated using the mode-summation or using the stress tensor approach. Blaming the mirrors in this case, then, may seem like special pleading. Nevertheless, curved boundaries are known to represent additional challenges in Casimir physics [11]. It remains a distinct possibility that the divergence at $r = a$ is an artifact of assuming a perfect spherical mirror, whereas a mirror modelled along more realistic lines might lead to a finite force.

As for the third possibility, it is clear that we abandoned Lifshitz' regularisation in our calculation—at least *formally*. Instead, we focussed on the term due to scattering at the mirror. Herein lies the conundrum: ordinarily, Lifshitz' regularisation is supposed to recover this term too, so the two approaches should coincide. But in this case they do not appear to. It is from the privileged perspective of the fish-eye's 'virtual space' that we are able to identify and isolate a reflection term, which is not identical to the term that remains from subtracting the auxiliary Green function of a homogeneous space. But what principles are at stake here? We have described the motivation behind the classic procedure as aimed at effecting a kind of renormalisation, in which the Casimir force is reset to zero for a homogeneous space. In this schema, contributions to the Casimir stress arise to the extent that the system differs from a flat homogeneous space. But perhaps this is not the best way to think about regularisation, and there is a different way to construe the basic procedure—one which recovers the correct method for extracting finite results for homogeneous systems, but does not misapply it beyond its proper domain of validity. Suppose, like Casimir, we were to construe the correct regularisation as essentially effecting an energy difference between the basic constituents at finite and at infinite separations in the system.[12] The basic constituents, in this case, are the features that make up the inhomogeneities in the system responsible for scattering electromagnetic radiation, and what remains from the calculation is the energy required to bring them from infinity into their present spatial configuration.

For the homogeneous case, the basic constituents assembled at finite separations is represented by the Green function. The auxiliary Green function, defining a homogeneous medium with the properties of the point of measurement, represents the state in which all the constituents around the locality of that point have been infinitely separated. It follows that Lifshitz' procedure for systems composed of homogeneous media is the correct one. However, this analysis may also offer some physical insight into what is going wrong with the inhomogeneous case. Here, the basic constituents (the inhomogeneities) have been brought *infinitesimally* close to each other, though it requires in fact an infinite amount of energy to achieve such an arrangement.

[12] This picture of regularisation is owed to a discussion with Simon Horsley, in which he suggested applying Casimir's interpretation of regularisation to Lifshitz theory too.

This introduces into the calculation an additional divergence that is untouched by Lifshitz' procedure of regularisation. On this understanding, it would appear that the correct method for calculating Casimir forces in inhomogeneous media involves an additional step—one in which the 'grainy' condition of the medium is properly acknowledged. Perhaps this is automatically achieved by incorporating spatial dispersion.

On this analysis, we also gain a fresh perspective on the fish-eye problem. In this case, we effectively took the difference between the Green's function of a fish-eye with a mirror and without one; the unphysical divergence associated with assembling the inhomogeneous material of the fish-eye was done away with—almost, but not quite. As we have observed, the sphere has curvature, which represents something of the inhomogeneity of the medium.[13] With a perfect mirror in place, the shorter wavelength modes are perhaps able to resolve the shape of the surface to arbitrary resolution, when in reality they should cease to 'see' the mirror at all beyond a certain lengthscale, and should therefore cease to contribute to the Casmir stress. How the mirror is itself 'assembled' as a structure, then, should perhaps also be taken into consideration. In short, it may be that the answer to this conundrum involves a superposition of all three of the possibilities first suggested.

## 7.5 Summary Remarks

The conundrum of the Casimir force in Maxwell's fish-eye challenges us to reconsider not only how the microphysical details of macroscopic media may feature in the computation of Casimir forces, but also the *meaning* of regularisation in the context of Lifshitz theory. It is clear that a Lifshitz-regularised stress tensor does not yield the correct prediction for the Casimir force in the fish-eye, either with a mirror or without one.

However, we have been able to determine an exact solution for the Casimir stress in an inhomogeneous medium by identifying and refocussing the calculation on the scattered part of the electromagnetic field. It is attractive rather than repulsive, and that in itself is surprising, given the almost iconic association of repulsive Casimir forces with spherical geometries. Analytic solutions for Casimir forces are extremely rare and this is the first exact solution for a non-uniform spherically symmetric medium. The stress, however, diverges at the boundary, so it is impossible to predict the Casimir force on the mirror. Arguably, the correct prediction of the force will depend on the properties of the mirror, and could lead to the prediction of an attractive vacuum force. If so, this could lend new hope to Casimir's fascinating explanation of the

[13] The inhomogeneous nature of the material cannot simply be removed by a coordinate transformation, as it was in Chap. 6.

finestructure constant [5] and the stability of elementary charged particles,[14] and settle once and for all the question of whether or not the Casimir force on a spherical shell is attractive or repulsive.

## References

1. U. Leonhardt, W.M.R. Simpson, Phys. Rev. D **84**, 081701 (2011)
2. U. Leonhardt, T.G. Philbin, *Geometry and Light: The Science of Invisibility* (Dover, New York, 2010)
3. T.G. Philbin, C. Xiong, U. Leonhardt, Ann. Phys. **325**, 579 (2009)
4. W.M.R. Simpson, S.A.R. Horsley, U. Leonhardt, Phys. Rev. A (2013)
5. H.B.G. Casimir, Physica (Utrecht) **19**(846) (1953)
6. J.D. Jackson, *Classical Electrodynamics* (Wiley, New York, 1998)
7. T.H. Boyer, Phys. Rev. **174**(1764) (1968)
8. K.A. Milton, L.L. DeRaad Jr., J. Schwinger, Ann. Phys. Ann. Phys. **115**(388) (1978)
9. S.A.R. Horsley, Phys. Rev. A **86**, 023830 (2012)
10. U. Leonhardt, T.G. Philbin, Phys. Rev. A **81**, 011804 (2010)
11. E.B. Kolomeisky, J.P. Straley, L.S. Langsjoen, H. Zaidi, J. Phys. A **43**, 385402 (2010)

[14] Of course, in a more realistic theory the particle should not be regarded as a classical object interacting with the quantum vacuum, as in Casimir's case, but rather as a self-consistent quantum structure.

# Chapter 8
# Outlook

*We see through a glass but darkly.*

St Paul, *1 Corinthians 13*

We are coming to the end of our discussion on the subject of quantum forces in inhomogeneous media, though there are doubtless many more 'surprises' in store in Casimir Physics even for the savant. It may be helpful, however, to review in brief some of the surprises and conundrums we have discussed, the interesting results that these questions gave rise to, and some possible ways ahead for answering them.

## 8.1 The Cut-Off Independence of the Casimir Energy

In Chap. 4, we extended Casimir's original calculation more generally, noting that, whilst Casimir forces between macroscopic bodies have been determined for a variety of systems and geometries [1, 2], most cases involve an idealisation in which the optical properties of the interacting media are perfectly homogeneous. Casimir's approach, detailed in Chap. 1, involves determining the ground-state energy of a system composed of two parallel plates,

$$E = \frac{\hbar}{2} \sum_{k} \rho(k)\, \omega_k, \tag{8.1.1}$$

and extracting a finite energy $E_{Casimir}$ that depends on the distance $a$ between them. This is achieved by applying a regulariser to the energy $E \rightarrow E(\xi)$, as described in Sect. 1.1.2, effectively separating the energy into two components[1] (for small $\xi$):

$$E(\xi) = E_{\infty}(\xi) + E(a), \tag{8.1.2}$$

[1] Properly, there is a third component $E(a, \xi)$ that contains mixed terms in $a$ and $\xi$, where the powers of $\xi$ are positive. This contribution is zero for $\xi = 0$, however.



W.M.R. Simpson, *Surprises in Theoretical Casimir Physics*, Springer Theses,
DOI 10.1007/978-3-319-09315-4_8

where $E_\infty(\xi)$ does not depend upon the distance between the plates. The cut-off $\xi$ is removed at the last step of the calculation:

$$E_{Casimir} = \lim_{\xi\to 0}\left[E(\xi) - E_\infty(\xi)\right]. \tag{8.1.3}$$

For the case of vacuum, or a homogeneous medium between the places, $E_{Casimir}$ is a finite quantity: $E_\infty$ diverges in the limit $\xi \to 0$, but $E(a)$ does not. Likewise, we discovered that, on introducing a gently inhomogeneous fluid between the plates, in which the optical properties of the medium vary continuously across the length of the cavity $\epsilon \to \epsilon(x)$, the ground-state energy of the system remains finite in the limit $\xi \to 0$, and consequently a finite mechanical energy, depending only on the distance $a$, can be stated exactly.

However, we should not hold any conclusions based on this thought-experiment too doggedly. Casimir's model is highly idealised, employing non–dispersive boundary conditions and considering only a simple energy summation of the field modes. On introducing the fluid into the model, we also made the artificial assumption that the electric permittivity is independent of frequency, in order to successfully compute the mode summation. Real media are both dispersive and dissipative. In order to account for these phenomena properly, and attempt to generalise our result beyond the highly artificial constraints of this model, we required a more sophisticated approach.

## 8.2 The Divergence of the Casimir Stress

In Chap. 5 we examined the local behaviour of the regularised stress tensor[2] commonly used in calculations of the Casimir force. As argued in Chap. 2, Casimir forces can be computed using a quantum analogue of the Minkowski stress tensor

$$\sigma_M = (\mathbf{D}\otimes\mathbf{E}) + (\mathbf{B}\otimes\mathbf{H}) - \frac{1}{2}(\mathbf{D}\cdot\mathbf{E} + \mathbf{B}\cdot\mathbf{H})\,\mathbf{1}_3, \tag{8.2.1}$$

which can be written in terms of the electromagnetic Green function describing the field produced by sources of current within the system, as described in Sect. 2.3.2, from which a finite stress can be extracted. In this case, this is achieved by a point-splitting of the stress $\sigma(x) \to \sigma(x, x')$, as described in Sect. 2.3.3, in which the stress is decomposed into the form

$$\sigma(x, x') = \sigma_\infty(x, x') + \sigma_C(x, x'), \tag{8.2.2}$$

[2] See Sect. 2.3.3.

where $\sigma_\infty(x, x')$ does not depend upon the local inhomogeneity of the medium, and diverges in the limit $x \to x'$. The two points are ultimately reunited at the last step:

$$\sigma_{Casimir}(x) = \lim_{x' \to x} \left[\sigma(x, x') - \sigma_\infty(x, x')\right]. \tag{8.2.3}$$

This approach is based on Lifshitz theory [3, 4]. For the case of a dielectric medium varying in one direction, the relevant component of the stress takes the form

$$\sigma_{Casimir}(x) = 2\hbar c \sum_{q=s,p} \int_0^\infty \frac{d\kappa}{2\pi} \int_{\mathbb{R}^2} \frac{d^2\boldsymbol{k}_\parallel}{(2\pi)^2} w \frac{r_{qL} r_{qR} e^{-2aw}}{1 - r_{qL} r_{qR} e^{-2aw}}, \tag{8.2.4}$$

expressed in terms of a set of reflection coefficients $r_{qL}$ for a small cavity, as described in Sect. 5.2. The advantage of using this approach is its capacity to incorporate dispersion and dissipation, and therefore to model real media. This represented a marked improvement on our first thought-experiment. For piece-wise homogeneous media, $\sigma_{Casimir}$ is finite. However, we found that the stress tensor was not finite anywhere within an inhomogeneous medium. On introducing a continuously varying medium, $\sigma_C$ decomposes into the form

$$\sigma_C = \tilde{\sigma}(\epsilon, \mu) + \tilde{\sigma}_\infty(\epsilon', \mu'), \tag{8.2.5}$$

where $\tilde{\sigma}_\infty$ is a contribution depending upon derivatives of the dielectric functions which diverges in the limit $x' \to x$. Since the Casimir force depends on inhomogeneities in the medium, as argued in Sect. 2.3, absorbing this contribution into $\sigma_\infty$ would constitute an additional renormalisation for which there is no general physical justification. Importantly, this result holds whatever the temporal dispersion, absorption properties or refractive index profile of the medium.

Conservatively, we can at least conclude that the Casimir stress tensor (5.2.1) for a piecewise definition of an inhomogeneous medium does not represent an approximation to the continuous case, and that the divergence of the stress tensor is not removed by the procedure of regularisation usually advocated. Further, as discussed in Sect. 5.4.1, we have been able to connect the divergence more precisely to the unphysical contribution of high wave numbers in the continuum limit—a problem that does not seem to be widely appreciated in the literature. A possible explanation for this anomaly is that the Casmir force does not in fact depend on such small-scale inhomogeneities as a continuously varying medium introduces, and an additional physical parameter is needed to allow for the 'grainy' nature of real media (cf. [5, 6]). As discussed in Sect. 5.4.3, the behaviour of the stress tensor in the limit of high wave numbers might seem to suggest that incorporating the spatially dispersive propertiesof media may be a possible solution to this problem,[3] in which the permittivities and permeabilities become functions of the wavenumber as well as the frequency:

$$\epsilon(x, \omega) \rightarrow \epsilon(x, \omega, k), \quad \mu(x, \omega) \rightarrow \mu(x, \omega, k). \tag{8.2.6}$$

But this would be rather surprising: how could it be that such small changes to the properties of the medium as we have been considering suddenly make the microscopic details relevant? It does seems odd that an arbitrarily small perturbation would require that kind of detailed information in order to diffuse the divergence. On the other hand, divergences can be found in the stress on the interface for two homogeneous plates of different refractive index that are touching, and an inhomogeneous medium might be conceived as many such slices piled together [7].

Deriving a plausible theoretical model of the spatially dispersive properties of media, however, is a far from trivial task, and not one we would wish to undertake without sufficient cause. Moreover, we did not need detailed information about the microphysical properties of the medium in order to determine a finite Casimir energy using the mode summation method. In order to consolidate our understanding of the problem we have uncovered in Casimir theory, we decided to examine two cases of inhomogeneous media with interesting properties, to see how they might fit with our findings, and found ourselves confronted with two conundrums.

## 8.3 The Casimir Force in a 'Compressive Medium'

First, in Chap. 6, we imagined introducing an inhomogeneous, anisotropic transformation medium into a cavity of the form

$$\epsilon(z, \omega) = \mu(z, \omega) = \begin{pmatrix} m(z, \omega)^{-1} & 0 & 0 \\ 0 & m(z, \omega)^{-1} & 0 \\ 0 & 0 & m(z, \omega) \end{pmatrix}. \tag{8.3.1}$$

We called this device the $C$-slice. This impedance-matched profile was specifically designed to implement a simple distortion of the laboratory coordinate system, effectively compressing or expanding the measure of space for light along one axis. At first, it seemed this *Gedankenexperiment* could contradict our earlier conclusions: since the inhomogeneous $C$-slice merely modifies the optical distance between the walls of an empty cavity, it seems we should be able to predict the Casimir force using Lifshitz theory. For a non-dispersive $C$-slice occupying the interval $x \in [a, b]$, the Casimir force between two plates separated by a distance $d$ is easily stated:

$$P = -\frac{\hbar c \pi^2}{240\left(d + \Delta(C_S^{-1} - 1)\right)^4}, \tag{8.3.2}$$

[3] This suggestion is developed further in [7], which draws upon the results of Chap. 5.

where

$$C_S = \frac{1}{\Delta} \int_a^b m(z, \omega)\,\mathrm{d}x, \quad \Delta = b - a. \tag{8.3.3}$$

The conundum emerges from our earlier discussion, in which the Casimir stress in an inhomogeneous medium was found to be cutoff-dependent. In fact, there is no contradiction here, though the case of the $C$-slice compels us to qualify the nature of the problem with inhomogeneous media, which pertains more specifically to the phenomenon of scattering within a medium: the stress diverges and the force is cutoff-dependent because the reflection coefficients fail to fall off fast enough in the limit of high wave numbers. However, in the case of the $C$-slice, we find that the reflection coefficients are precisely zero for all angles of incidence, and the stress does not diverge. Such 'transformation media' merely change the measure of optical distance, and do not introduce additional scattering into the cavity. Therefore, it is possible to determine the Casimir forces for systems incorporating transformation media using Lifshitz theory, even when they involve continuously changing dielectric properties.

As a result of exploring this problem, however, we were able to determine and state exactly a finite Casimir force for an inhomogeneous medium. Such solutions are rare in the literature. Moreover, the $C$-slice has some interesting properties. A suitably configured $C$-slice interposed between two parallel plates can reduce the Casimir force between the plates arbitarily by increasing the effective size of the cavity. Such a medium could be applied to the problem of quantum stiction. However, the engineering difficulties in achieving the physical properties of the $C$-slice for a sufficiently broad range of the electromagnetic spectrum are extreme.

## 8.4 The Casimir Force in Maxwell's Fish-Eye

Finally, in Chap. 7, we considered the case of the Casimir force in Maxwell's fish-eye—an inhomogeneous metamaterial with an electric permittivity and magnetic permeability equal to

$$\varepsilon = \mu = n = \frac{2n_1}{1 + (r/a)^2}, \tag{8.4.1}$$

whose effect on light can be interpreted through the stereographic projection of geodesics on the surface of the four dimensional hypersphere, as discussed in Sect. 7.2.1. The hyperspherical geometry of the bare fish-eye would appear to preclude the possibility of a Casimir force, but the stress tensor (calculated in the usual way) is paradoxically infinite, even after regularisation.

However, by examining the problem of the fish-eye in a spherical mirror of radius $a$, in which we were able to identify and refocus the calculation on the scattered part of the field in the geometry of the hypershere, we were able to determine an exact solution for the Casimir stress in an inhomogeneous medium. The Green function for the total system decomposes into the form

$$G = G_i(r) - G_r(r), \tag{8.4.2}$$

where $G_i$ is the Green function of an incident wave and $G_r$ is an image source, and both waves are travelling on the hypersphere. Our method was to subtract the Green function correponding to a source on the bare hypersphere, which is equal to $G_i$, and compute the stress tensor using the remaining term $-G_r(r)$:

$$\sigma(r) = -\frac{\hbar c}{\pi^2 a^4 n\,(1 - r^2/a^2)^4}\mathbf{1}_3. \tag{8.4.3}$$

The stress is attractive. Given the association of repulsive Casimir forces with spherical geometries this seems rather surprising. The stress, however, diverges at the boundary $r \to a$, so it is still not possible to predict the size of the Casimir force on a spherical mirror. Any claims to date to have done so should perhaps be treated more sceptically. It seems likely that the correct prediction of the Casimir force on a spherical mirror depends (perhaps in detail) on the properties of the mirror, but that the force itself remains attractive. If so, this could have interesting implications for the stability of charged particles and the origin of the finestructure constant [8].

Nevertheless, the conundrum of the Casimir stress in Maxwell's fish-eye resides in an unresolved tension between the modified regularisation procedure, involving the subtraction of a Green function corresponding to a source on the hypersphere, and the standard procedure of regularisation within Lifshitz theory, involving the subtraction of an auxiliary Green function corresponding to an infinite homogeneous space. Dissolving this conundrum satisfactorily constitutes a challenge for future research in this area.

## 8.5 Beyond the Diverging Casimir Force

Let us end on an empirical note. Whilst Casimir-Lifshitz forces are notoriously difficult to measure, and the number of experiments investigating Casimir interactions between solid bodies is fairly small, the experimental field for Casimir-Polder forces is comparatively rich, and one in which subtle measurements can be made.[4] A possible generalisation of the work we have discussed in the last few chapters could involve the investigation of Casimir-Polder forces in inhomogeneous media. Does the divergence we have discovered in the Casimir force also affect the interaction between a solid body and a particle? Either way, the answer to this question seems

[4] See Sect. 1.2.2.

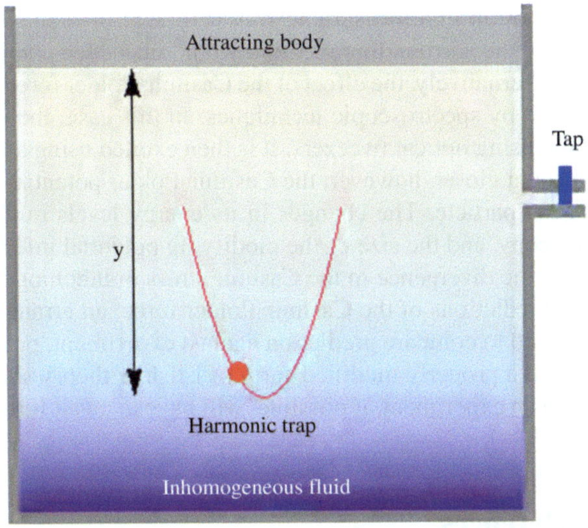


**Fig. 8.1** A chamber is filled with an inhomogeneous fluid (the fluid is becoming denser towards the bottom). A particle (*red*) is confined within the chamber in a harmonic trap by a laser. A body, sitting on the surface of the fluid, is brought steadily closer to the particle by siphoning off the fluid with a tap, modifying the motion of the particle in the trap through the Casimir-Polder interaction

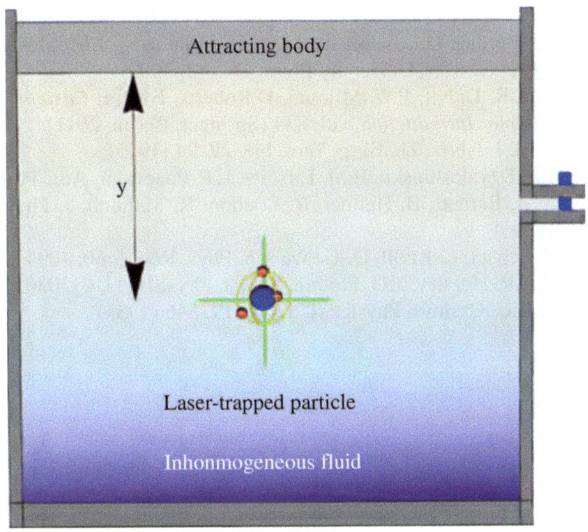


**Fig. 8.2** This setup is identical to Fig. 8.1 except the particle is now held fixed by optical tweezers and excited by a laser. The changes in its energy levels as the attracting body is brought closer are measured using spectroscopic techniques

likely to illuminate further the problems we have been considering. Moreover, such a project may be amendable to experimentation:

Imagine the following scenario: it is possible to optically confine a particle in a liquid using a laser. Suppose the liquid is an inhomogeneous fluid (for example, sugar dissolved in water, under gravity). Suppose further that we introduce a mirror that makes contact with the surface of the water, and that the fluid can be drained to different levels with a siphon, so that the distance between the reflective body and the confined particle is a variable we can control. The effect of the mirror on the particle could be studied using a confinement in which the particle undergoes oscillatory motion. The Casimir-Polder force would change the trapping potential, modifying the motion in a measurable way (Fig. 8.1). It seems likely that the viscious forces due to the surrounding medium would introduce complications, however.

Alternatively, the effect of the Casimir-Polder force on the particle could be ascertained by spectroscopic techniques. In this case, the particle is held in place in the fluid using optical tweezers. It is then excited using a laser. As the attracting body is brought closer, however, the Casimir-Polder potential will modify the energy levels of the particle. The changes in its energy levels might be ascertained using spectroscopy, and the size of the modifying potential inferred (Fig. 8.2).

If the divergence in the Casimir stress in inhomogeneous media similarly infects the predictions of the Casimir-Polder force, an arrangement like one of these could be used to compare prediction against experiment, providing the finite curve against which a properly modified form of Lifshitz theory should be made to fit. However, such an experiment, if possible, will have to await further theoretical developments.

## References

1. M. Bordag, G.L. Klimchitskaya, U. Mohideen, V.M. Mostepanenko, *Advances in the Casimir Effect* (Oxford University Press, Oxford, 2009)
2. D.A.R. Dalvit, P.W. Milonni, D.Roberts, F.Rosa, *Casimir Physics: Lecture Notes in Physics. Chapter Introduction*, vol. 834 (Springer, Berlin, 2011)
3. E.M. Lifshitz, Zh. Eksp. Teor. Fiz. **29**, 94 (1955)
4. I.E. Dzyaloshinskii, E.M. Lifshitz, L.P. Pitaevskii, Adv. Phys. **10**, 165 (1961)
5. S.M. Barnett, B. Huttner, R. Loudon, R. Matloob, J. Phys. B: At. Mol. Opt. Phys. **29**, 3763 (1996)
6. S. Scheel, L. Knöll, D.-G. Welsch, Phys. Rev. A **60**, 4094 (1999)
7. S.A.R. Horsley, T.G. Philbin, New J. Phys. **16**(1), 013030 (2014)
8. H.B.G. Casimir, Physica (Utrecht) **19**, 846 (1953)

# Appendix A
# Causality and Wick Rotation

## A.1 Concerning Causality

Consider a physical system[1] perturbed by an input $I$, resulting in a response $R$ at time $t$. We can relate the input and the response via a linear response function $G$ as follows:

$$R(t) = \frac{1}{\sqrt{2\pi}} \int_{-\infty}^{\infty} G(t - t')J(t')\, \mathrm{d}t'. \tag{A.1}$$

For example, $J(t')$ could be a current source at time $t'$ producing an electromagnetic field measured at $t$, where $G$ is the Green function of the electromagnetic field. We assume that $G$ depends only on $t - t'$; the system under consideration should respond to a sharp input at $t = t_0$ in the same way as it would respond to the input at a subsequent time $t = t_0 + \tau$:

$$R(t + \tau) = R(t). \tag{A.2}$$

A physical system that receives an input at time $t$ cannot respond at times prior to $t$. That is, $G(\tau) = 0$ for $\tau < 0$. This is known as the *causality requirement*, and it may be expressed in the form

$$R(t) = \frac{1}{\sqrt{2\pi}} \int_{-\infty}^{t} G(t - t')J(t')\mathrm{d}t', \tag{A.3}$$

in which the response at $t$ is written as a weighted linear superposition of all responses prior to $t$. We may also assume that a physical system at time $t$ is not affected by an input in the remote past, i.e. that $G(\tau) \to 0$ as $\tau \to \infty$. This may be construed as the

[1] This section is indebted to the exposition of the theory of analytic functions in [1].



W.M.R. Simpson, *Surprises in Theoretical Casimir Physics*, Springer Theses,
DOI 10.1007/978-3-319-09315-4

requirement for some mechanism of dissipation in a physical system that dampens the response to an impulse so that it eventually becomes negligible:

$$\int_0^\infty |G(\tau)| \mathrm{d}\tau = M < \infty, \tag{A.4}$$

where $M$ is the finite upper-bound of the integral. These physical assumptions about the system may be summarised in the form of three conditions on $G$:

i $G(\tau)$ is bounded for all $\tau$.
ii $|G(\tau)|$ is integrable, so $G(\tau) \to 0$ faster than $1/\tau$ as $\tau \to \infty$.
iii $G(\tau) = 0$ for $\tau < 0$.

In the Fourier transform,[2] Eq. (A.1) can be rewritten as

$$R(\omega) = G(\omega) J(\omega). \tag{A.5}$$

It follows from (i) and (ii) that both $G(\tau)$ and $G(\omega)$ are square integrable. In electromagnetism, the presence of dissipation in a system typically manifests in the form of an imaginary component (for example, in the electric permittivity). We must now extend $G(\omega)$ to the complex plane. Recalling that $G(\tau) = 0$ for $\tau < 0$, we define

$$G(z) = \frac{1}{\sqrt{2\pi}} \int_0^\infty G(t) e^{izt} \mathrm{d}t = \frac{1}{\sqrt{2\pi}} \int_0^\infty G(t) e^{i\omega t} e^{-\omega' t} \mathrm{d}t, \tag{A.6}$$

where $z = \omega + i\omega'$. The causality requirement (iii) implies that the term $e^{-\omega' t}$ is a decaying exponential in the upper half-plane ($\omega' > 0$). Restating $z$ in the form $|z| e^{i\phi} \implies \omega' = |z| \sin\phi$, we deduce that

$$|g(z)| \leq \frac{1}{\sqrt{2\pi}} M \int_0^\infty \mathrm{d}t \, \mathrm{e}^{-(|z| \sin\phi) t} = \frac{M}{\sqrt{2\pi} \, |z| \sin\phi}, \tag{A.7}$$

in the interval $0 < \phi < \pi$, where we have invoked assumption (A.4). Clearly, this expression tends to zero in the limit $|z| \to \infty$. For $\phi = 0$ or $\pi$, we obtain $\omega' = 0$, leaving $G(\omega)$, which is square integrable. It follows that, in any direction in the upper

[2] Two quantities $F(\omega)$ and $f(t)$ are Fourier transform pairs when they are connected by the Fourier transform

$$F(\omega) = \frac{1}{\sqrt{2\pi}} \int_{-\infty}^{\infty} f(t) e^{i\omega t} \mathrm{d}t.$$

half-plane, $|G(z)| \to 0$ as $|z| \to \infty$. It is also clearly apparent, on this representation of $G(z)$, that the function is analytic in the upper half-plane[3] ($\omega' > 0$):

$$\frac{d^n G}{dz^n} = \frac{1}{\sqrt{2\pi}} \int_0^\infty G(t) \frac{d^n}{dz^n} e^{izt} dt = \frac{i^n}{\sqrt{2\pi}} \int_0^\infty t^n G(t) e^{i\omega t} e^{-\omega' t} dt. \tag{A.8}$$

In summary, for any $G(z)$ arising from $G(t)$ that satisfies our original assumptions (i), (ii) and (iii), it follows that

i. $|G(z)| \to 0$ as $|z| \to \infty$.
ii. $G(z)$ is analytic in the upper half-plane ($\omega' > 0$).

These are also the conditions which, if satisfied, induce a Hilbert transform pair relating the real and imaginary parts of a complex function [1]. It follows that

$$\text{Re}\, G(\omega) = \frac{1}{\pi} P \int_{-\infty}^{\infty} \frac{\text{Im}\, G(\omega')}{\omega' - \omega} d\omega', \tag{A.9}$$

$$\text{Im}\, G(\omega) = -\frac{1}{\pi} P \int_{-\infty}^{\infty} \frac{\text{Re}\, G(\omega')}{\omega' - \omega} d\omega', \tag{A.10}$$

where $P$ denotes the principal-value integral.[4] Since only positive frequencies have empirical meaning in this context, these relations should be rewritten in a more useful form. From (A.6), we deduce the so-called *reality condition*:

$$G^*(z) = G(-z^*), \tag{A.11}$$

having noted that $G(t)$ is real. If $z$ is real, i.e. $z = \omega$, by comparing the real and imaginary parts of (A.11), we deduce that

$$\text{Re}\, G(\omega) = \text{Re}\, G(-\omega) \tag{A.12}$$

$$\text{Im}\, G(\omega) = -\text{Im}\, G(-\omega), \tag{A.13}$$

i.e. $\text{Re}\, G(\omega)$ is an even function, and $\text{Im}\, G(\omega)$ is odd. Using equation (A.16), split into two integrals with limits $\{-\infty, 0\}$ and $\{0, \infty\}$, and substituting $\omega' \to -\omega'$ in

[3] These arguments do not exclude the possibility of a singularity at $\omega' = 0$. However, the Hilbert transform is tolerant of bounded branch point singularities on the real axis [1].

[4] The principal-value integral of a function $f(x)$ is defined

$$P \int_{-x'}^{x'} f(x)\, dx = \int_{-x'}^{\alpha-\delta} f(x)\, dx + \int_{\alpha+\delta}^{x'} f(x)\, dx.$$

the first integral, we obtain

$$\mathrm{Re}\, G(\omega) = \frac{1}{\pi} P \int_{\infty}^{0} \frac{\mathrm{Im}\, G(-\omega')}{\omega' - \omega} \mathrm{d}\omega' + \frac{1}{\pi} P \int_{-\infty}^{0} \frac{\mathrm{Im}\, G(\omega')}{\omega' - \omega} \mathrm{d}\omega'. \tag{A.14}$$

Using the relation (A.13), we can rewrite the above in the compact form

$$\mathrm{Re}\, G(\omega) = \frac{2}{\pi} P \int_{0}^{\infty} \frac{\omega' \mathrm{Im}\, G(\omega')}{\omega'^2 - \omega^2} \mathrm{d}\omega'. \tag{A.15}$$

Similarly, using Eqs. (A.10) and (A.12), we obtain

$$\mathrm{Im}\, G(\omega) = -\frac{2\omega}{\pi} P \int_{0}^{\infty} \frac{\mathrm{Re}\, G(\omega')}{\omega'^2 - \omega^2} \mathrm{d}\omega'. \tag{A.16}$$

These are the *Kramers–Kronig relations*, which connect the real and imaginary parts of the physical quantity $G(\omega)$ together for real values of the argument $\omega$ by a dispersion relation. Since the imaginary part of a response function typically describes how energy is dissipated in a physical system, the Kramers–Kronig relations imply that observing the dissipative response of a system is sufficient to determine its dispersive effects, and vice versa.

## A.2 Wick Rotation

The Casimir Effect is a broadband phenomenon, and the computation of Casimir forces typically involves the integration of quantities over a wide range of frequencies. These integrals are oscillatory and slow to converge. However, using contour-integral techniques, it is possible to transform real-frequency integrals into integrals along the positive imaginary axis, resulting in more rapid convergence. Consider the quantity

$$\int_{0}^{\infty} \mathrm{Im}\, G(\omega)\, \mathrm{d}\omega = \frac{1}{2i} \int_{0}^{\infty} \mathrm{d}\omega \left[ G(\omega) - G^{*}(\omega) \right]. \tag{A.17}$$

Recalling (A.11), we can rewrite this integral as

$$\frac{1}{2i} \int_{0}^{\infty} \mathrm{d}\omega \left[ G(\omega) - G(-\omega) \right] = \frac{1}{2i} \int_{0}^{\infty} \mathrm{d}\omega\, G(\omega) + \frac{1}{2i} \int_{0}^{-\infty} \mathrm{d}\omega\, G(\omega). \tag{A.18}$$

As a consequence of the causality of the physical quantities under consideration (for example, electromagnetic fields produced by fluctuations in the material arise *after* the source currents are generated, not before), we conclude that they are analytic functions in the upper half of the complex frequency plane. Cauchy's theorem requires that any closed integral in the upper part of the complex plane must vanish:

$$\oint G(\omega)\,\mathrm{d}\omega = 0. \tag{A.19}$$

This analyticity implies than our integral can be arbitrarily deformed to any contour in this part of the complex plane. Setting $\omega = i\xi$, $\mathrm{d}\omega = i\mathrm{d}\xi$, with real $\xi$ running over the positive imaginary axis, and introducing polar coordinates $\omega = |\omega|e^{i\phi}$, we may write

$$\int_0^\infty G(\omega)\,\mathrm{d}\omega + \lim_{|\omega|\to\infty} \int_0^{\pi/2} i\omega\,\mathrm{d}\phi\, G(\omega) + \int_\infty^0 i\mathrm{d}\xi\, G(i\xi) = 0, \tag{A.20}$$

where we follow a contour composed of an integration along the positive real axis, an integration over a quarter-circle to the imaginary axis, and a final integration down the imaginary axis, closing the contour. It follows that

$$\int_0^\infty G(\omega)\,\mathrm{d}\omega = i\int_0^\infty \mathrm{d}\xi\, G(i\xi) - \lim_{|\omega|\to\infty} i\int_0^{\pi/2} \mathrm{d}\phi\,\omega G(\omega). \tag{A.21}$$

Similarly, we find that

$$\int_0^{-\infty} G(\omega)\,\mathrm{d}\omega = i\int_0^\infty \mathrm{d}\xi\, G(i\xi) + \lim_{|\omega|\to\infty} i\int_{\pi/2}^{\pi} \mathrm{d}\phi\,\omega G(\omega). \tag{A.22}$$

In the limit $|\omega| \to \infty$, the integrals along the infinite quarter-circles vanish. We thus obtain the useful transformation

$$\int_0^\infty \mathrm{d}\omega\,\mathrm{Im}G(\omega) = \int_0^\infty \mathrm{d}\xi\, G(i\xi). \tag{A.23}$$

## Reference

1. F.W. Byron, R.W. Fuller, *Mathematics of Classical and Quantum Physics* (Dover, New York, 1992)

# Appendix B
# Maxwell's Stress Tensor

Consider the expression for the Lorentz force density:

$$\mathbf{f} = \rho\mathbf{E} + \mathbf{J} \times \mathbf{B}. \tag{B.1}$$

For simple problems, calculating the force on a point charge using the Lorentz force law is straightforward. For more general and more complicated cases, this becomes increasingly difficult. We can eliminate the density $\rho$ and the current $\mathbf{J}$ by using Maxwell's equations:

$$\nabla.\mathbf{E} = \frac{\rho}{\varepsilon_0}, \quad \nabla.\mathbf{B} = 0, \quad \nabla \times \mathbf{E} = -\frac{\partial \mathbf{B}}{\partial t}, \quad \nabla \times \mathbf{B} = \mu_0 \mathbf{J} + \mu_0 \varepsilon_0 \frac{\partial \mathbf{E}}{\partial t}, \tag{B.2}$$

rewriting the Lorentz force solely in terms of the fields:

$$\mathbf{f} = \varepsilon_0 (\nabla.\mathbf{E})\mathbf{E} + \frac{1}{\mu_0}(\nabla \times \mathbf{B}) \times \mathbf{B} - \varepsilon_0 \frac{\partial \mathbf{E}}{\partial t} \times \mathbf{B}. \tag{B.3}$$

The time derivative can be rewritten in terms of the Poynting vector,

$$\begin{aligned} \frac{\partial}{\partial t}(\mathbf{E} \times \mathbf{B}) &= \frac{\partial \mathbf{E}}{\partial t} \times \mathbf{B} + \mathbf{E} \times \frac{\partial \mathbf{B}}{\partial t} \\ &= \frac{\partial \mathbf{E}}{\partial t} \times \mathbf{B} - \mathbf{E} \times (\nabla \times \mathbf{E}). \end{aligned} \tag{B.4}$$

It follows that

$$\frac{\partial \mathbf{E}}{\partial t} \times \mathbf{B} = \frac{\partial}{\partial t}(\mathbf{E} \times \mathbf{B}) + \mathbf{E} \times (\nabla \times \mathbf{E}). \tag{B.5}$$

We may then write

$$\mathbf{f} = \varepsilon_0 (\nabla.\mathbf{E})\mathbf{E} + \frac{1}{\mu_0}(\nabla \times \mathbf{B}) \times \mathbf{B} - \varepsilon_0 \frac{\partial}{\partial t}(\mathbf{E} \times \mathbf{B}) - \varepsilon_0 \mathbf{E} \times (\nabla \times \mathbf{E}). \tag{B.6}$$



W.M.R. Simpson, *Surprises in Theoretical Casimir Physics*, Springer Theses,
DOI 10.1007/978-3-319-09315-4

Collecting the **B** and **E** terms, introducing the zero-term $\frac{1}{\mu_0}(\nabla.\mathbf{B})\mathbf{B} = 0$ and utilising the cross product rule, we obtain

$$\mathbf{f} = \varepsilon_0\left((\nabla.\mathbf{E})\mathbf{E} - \mathbf{E}\times(\nabla\times\mathbf{E})\right) + \frac{1}{\mu_0}\left((\nabla.\mathbf{B})\mathbf{B} - \mathbf{B}\times(\nabla\times\mathbf{B})\right) - \varepsilon_0\frac{\partial}{\partial t}(\mathbf{E}\times\mathbf{B}). \tag{B.7}$$

The double-curls can be replaced using the curl identity

$$\frac{1}{2}\nabla A^2 = \mathbf{A}\times(\nabla\times\mathbf{A}) + (\mathbf{A}\cdot\nabla)\mathbf{A} \tag{B.8}$$

to give

$$\mathbf{f} = \varepsilon_0\left((\nabla.\mathbf{E})\mathbf{E} + (\mathbf{E}\cdot\nabla)\mathbf{E}\right) + \frac{1}{\mu_0}\left((\nabla.\mathbf{B})\mathbf{B} + (\mathbf{B}\cdot\nabla)\mathbf{B}\right)$$
$$-\frac{1}{2}\nabla\left(\varepsilon_0 E^2 + \frac{1}{\mu_0}B^2\right) - \varepsilon_0\frac{\partial}{\partial t}(\mathbf{E}\times\mathbf{B}).$$

However, this is a cumbersome expression. It is more convenient to collect these terms within a single tensor:

$$\hat{\sigma} = \varepsilon_0\mathbf{E}\otimes\mathbf{E} + \frac{1}{\mu_0}\mathbf{B}\otimes\mathbf{B} - \frac{1}{2}\left(\varepsilon_0\mathbf{E}\cdot\mathbf{E} + \frac{1}{\mu_0}\mathbf{B}\cdot\mathbf{B}\right)\mathbf{1}_3. \tag{B.9}$$

Consequently, the force can be re-expressed in terms of the stress $\hat{\sigma}$ and the Poynting vector **S**,

$$\mathbf{f} + \varepsilon_0\frac{\partial}{\partial t}\mathbf{S} = \nabla\cdot\hat{\sigma}, \quad \mathbf{S} = \frac{1}{\mu_0}(\mathbf{E}\times\mathbf{B}). \tag{B.10}$$

This stress tensor is known as the Maxwell stress tensor, which governs the flow of momentum associated with the electromagnetic field.

# Appendix C
# Green Functions from Transfer Matrices

The Casmir stress can be written in terms of reflection coefficients (2.4.26), or equivalently in terms of a Green function (2.3.75). We demonstrate here that the Green function derived from the transfer matrices (involving a piece-wise approximation of the media into homogeneous slices) recovers the same numerics as the Green function obtained from a numerical solution of the wave equation. The two analytic formulations of the stress above are identical. It follows that the piece-wise approximation (at a suitably high resolution) recovers the same numerics as the stress computed using a Green function obtained from a numerical solver for a continuous refractive index profile $n(x)$.

## C.1 Solving the Wave Equation Numerically

We are considering a planar geometry in which the media is inhomogeneous in $x$. To his end, we need only concern ourselves here with finding the scalar Green function $g(x)$, which is the solution of the wave equation [1]:

$$\partial_x \frac{1}{n(x)} \partial_x \, g(x, x_0) - \left( \frac{u^2 + v^2}{n(x)} + \kappa^2 n(x) \right) g(x, x_0) = \delta(x - x_0). \tag{C.1}$$

For simplicity, the medium is impedance-matched: $\epsilon(x) = \mu(x)$. It is difficult to solve the equation numerically in this form, however, as standard numerical solvers do not know how to 'handle' the dirac-delta function. However, this difficulty can be easily circumvented. The scalar Green function $g(x, x_0)$ can be rewritten in the following form:

$$g_N(x, x_0) = \begin{cases} c_1 g_l(x) & x \leq x_0, \\ c_2 g_r(x) & x > x_0, \end{cases} \tag{C.2}$$



W.M.R. Simpson, *Surprises in Theoretical Casimir Physics*, Springer Theses,
DOI 10.1007/978-3-319-09315-4

where $g_l(x)$ and $g_r(x)$ solve the equation

$$\partial_x \frac{1}{n(x)} \partial_x\, g_{l,r}(x) - \left( \frac{u^2 + v^2}{n(x)} + \kappa^2 n(x) \right) g_{l,r}(x) = 0, \tag{C.3}$$

for $x < x_0$ and $x > x_0$ respectively. By solving the differential equation for $g_l(x)$ and $g_r(x)$ separately, away from the source point, the dirac-delta function disappears from the differential equation. We obtain $g(x, x_0)$ by piecing both solutions together (C.2) in a way that satisfies the original differential equation (C.1), i.e. by choosing the correct values for the coefficients $c_1$ and $c_2$. The coefficients can be uniquely specified by two simultaneous equations. First, we require that

$$c_1 g_l(x) = c_2 g_r(x), \tag{C.4}$$

i.e. that the two solutions (to the left and right of the source point) should be identical at the source point. The second condition on the coefficients should recover the delta-function. Consider again the equation

$$\partial_x \frac{1}{n(x)} \partial_x\, g(x) - \left( \frac{u^2 + v^2}{n(x)} + \kappa^2 n(x) \right) g(x) = \delta(x - x'). \tag{C.5}$$

Integrating both sides with respect to $x$, between the limits of the source point and the point of measurement, we obtain

$$\int_x^{x_0} \mathrm{d}x' \left[ \partial_{x'} \frac{1}{n(x')} \partial_{x'}\, g(x') \right] - \int_x^{x_0} \mathrm{d}x' \left( \frac{u^2 + v^2}{n(x')} + \kappa^2 n(x') \right) g(x') = 1. \tag{C.6}$$

In the limit as $x \to x_0$, the second integral tends to zero, and we obtain

$$\lim_{x \to x_0} \left[ \frac{1}{n(x)} \partial_x\, g(x) - \frac{1}{n(x_0)} \partial_x\, g(x)|_{x \to x_0} \right] = 1. \tag{C.7}$$

Thus we arrive at the second condition on the coefficients:

$$[c_1 \partial_x g_l(x) - c_2 \partial_x g_r(x)]_{x \to x_0} = n(x_0). \tag{C.8}$$

These two conditions together determine the effect of the source. Suppose we introduce an inhomogeneous medium between $x = x_1$ and $x = x_2$. For $x < x_1$ and $x > x_2$, however, there is only vacuum. In these outer regions, the wave equation (C.1) is easily solvable, having solutions of the form

$$A e^{wx} + B e^{-wx}, \tag{C.9}$$

where $w = \sqrt{u^2 + v^2 + n(x)\kappa^2}$. For $x < x_1$, there is no reflection from the left, and for $x > x_2$, here is no reflection from the right, so we may impose the boundary conditions

$$g_l(x_1) = 1, \quad \partial_x g_l(x_1) = w(x_1), \tag{C.10}$$

$$g_r(x_2) = 1, \quad \partial_x g_r(x_2) = -w(x_2). \tag{C.11}$$

The magnitude of $g_l(x_1)$ and $g_r(x_2)$ may be fixed arbitrarily; it is the ratio of the functions and their derivatives at $x_1$ and $x_2$ that is important. Thus equation (C.3) can be solved with the boundary conditions (C.10) and (C.11), and equations (C.4) and (C.8) can be solved simultaneously to determine the constants $c_1$ and $c_2$, uniquely determining the scalar Green function $g(x, x_0)$ via (C.2).

## C.2 Green Function from Transfer Matrices

Here, we demonstrate how to recover the Green function using a piecewise approximation of the inhomogeneous region $(x_1, x_2)$ and applying transfer matrices between the pieces to determine the field in each slice. We divide the medium between $x_1$ and $x_2$ into $t$ slices of width $a = (x_2 - x_1)/t$. The left-hand side of each slice is positioned at $X_i$:

$$X_i = X(i) = \frac{i-1}{t-1}(x_2 - x_1) + x_1 - \frac{a}{2}, \tag{C.12}$$

and each slide is characterised by a constant refractive index

$$N(i) = n\left(X(i) + \frac{a}{2}\right), \tag{C.13}$$

where $n(x)$ is the refractive index profile of the system, so that the first slide is centred on $X_1 + a/2 = x_1$, with constant refractive index $N(1) = n(X(1) + a/2) = n(x_1)$, and the last slide is centred on $X_t + a/2 = x_2$, with constant refractive index $N(t) = n(X(t) + a/2) = n(x_2)$. Consider a cavity between $X_i$ and $X_{i+1}$. We will number this cavity $i$. Let us consider the field in cavity $i$. Sweeping from left to right, the field terms in the adjacent cell $i + 1$ can be stated in terms of the field terms in cell $i$. At the interface, we require that

$$E_i^{(R+)} + E_i^{(R-)} = E_{i+1}^{(L+)} + E_{i+1}^{(L-)}, \tag{C.14}$$

$$\frac{w_i}{n_i} E_i^{(R+) w_i a} - \frac{w_i}{n_i} E_i^{(R+) - w_i a} = \frac{w_{i+1}}{n_{i+1}} E_{i+1}^{(L+)} - \frac{w_{i+1}}{n_{i+1}} E_{i+1}^{(L-)}, \tag{C.15}$$

where $E_i^{(R)} = E_i^{(R+)} + E_i^{(R-)}$ is the field at the right of cell $i$, consisting of left and right wave amplitudes, $E_{i+1}^{(L)} = E_{i+1}^{(L+)} + E_{i+1}^{(L-)}$ is the field at the left of cell

$i + 1$, which also consists of left and right wave amplitudes, and $w_i = w(i) = \left(u^2 + v^2 + N(i)^2\kappa^2\right)^{1/2}$ is a Fourier-transformed wave number. This may be written more succinctly as

$$\begin{aligned}\begin{pmatrix} E_{i+1}^{(L+)} \\ E_{i+1}^{(L-)} \end{pmatrix} &= \frac{1}{2}\begin{pmatrix} \left(1 + \frac{w_i}{w_{i+1}}\frac{n_{i+1}}{n_i}\right) & \left(1 - \frac{w_i}{w_{i+1}}\frac{n_{i+1}}{n_i}\right) \\ \left(1 - \frac{w_i}{w_{i+1}}\frac{n_{i+1}}{n_i}\right) & \left(1 + \frac{w_i}{w_{i+1}}\frac{n_{i+1}}{n_i}\right) \end{pmatrix}\begin{pmatrix} E_i^{(R+)} \\ E_i^{(R-)} \end{pmatrix} \\ &= t_{i\to i+1}\begin{pmatrix} E_i^{(R+)} \\ E_i^{(R-)} \end{pmatrix}. \end{aligned} \tag{C.16}$$

To account for the evolution of the field, we introduce the phase term

$$\Lambda_{i+1} = \begin{pmatrix} \mathrm{e}^{-w_{i+1}a} & 0 \\ 0 & \mathrm{e}^{w_{i+1}a} \end{pmatrix}. \tag{C.17}$$

Then the combined transfer matrix becomes

$$T_{i\to i+1}^{R} = \Lambda_{i+1}t_{i\to i+1} = \frac{1}{2}\begin{pmatrix} \left(1 + \frac{w_i}{w_{i+1}}\frac{n_{i+1}}{n_i}\right)\mathrm{e}^{-w_{i+1}a} & \left(1 - \frac{w_i}{w_{i+1}}\frac{n_{i+1}}{n_i}\right)\mathrm{e}^{-w_{i+1}a} \\ \left(1 - \frac{w_i}{w_{i+1}}\frac{n_{i+1}}{n_i}\right)\mathrm{e}^{w_{i+1}a} & \left(1 + \frac{w_i}{w_{i+1}}\frac{n_{i+1}}{n_i}\right)\mathrm{e}^{w_{i+1}a} \end{pmatrix}, \tag{C.18}$$

which gives the field on the far right-hand side of a cell $i + 1$ in terms of the field on the far right-hand side of a cell $i$, i.e.

$$\begin{pmatrix} E_{i+1}^{(R+)} \\ E_{i+1}^{(R-)} \end{pmatrix} = T_{i\to i+1}^{R}\begin{pmatrix} E_i^{(R+)} \\ E_i^{(R-)} \end{pmatrix}. \tag{C.19}$$

By induction, the field at the far right-hand side of cell $k$ can be expressed in terms of the field at the far right-hand side of cell $j$, by successively applying the transfer matrices inbetween:

$$\begin{aligned}\begin{pmatrix} E_k^{(R+)} \\ E_k^{(R-)} \end{pmatrix} &= T_{k-1\to k}^{R}\cdots T_{i\to i+1}^{R}\begin{pmatrix} E_j^{(R+)} \\ E_j^{(R-)} \end{pmatrix} \\ &= \prod_{i=j}^{k} T_{i\to i+1}^{R}\begin{pmatrix} E_j^{(R+)} \\ E_j^{(R-)} \end{pmatrix} = \mathbb{T}_{j\to k}^{R}\begin{pmatrix} E_j^{(R+)} \\ E_j^{(R-)} \end{pmatrix}. \end{aligned} \tag{C.20}$$

Using this formalism, we will now recover the scalar Green function $g(x, x_0)$ for a source at $x = x_0$. We will label the cell in which the source has been placed with $i_0$.

Consider the field emerging into vacuum. According to (C.11), a right-going field should emerge at $t$; we will assign it a magnitude of unity. It follows that:

$$\begin{aligned}\begin{pmatrix}1\\0\end{pmatrix} &= \prod_{i=j}^{t-1} T^R_{i\to i+1}\begin{pmatrix}E_i^{(R+)}\\E_i^{(R-)}\end{pmatrix}\\ &= T^R_{t-1\to t}\cdots T^R_{i_0+1\to i_0+2}\cdot T^R_{j\to j+1}\begin{pmatrix}E_j^{(R+)}\\E_j^{(R-)}\end{pmatrix}\\ &= \mathbb{T}^R_j\begin{pmatrix}E_j^{(R+)}\\E_j^{(R-)}\end{pmatrix}.\end{aligned} \tag{C.21}$$

The successive multiplication terminates at $t-1$, as the final term $T_{t-1\to t}$ evolves the field up to the left-boundary of cell $t$, which has the form given above. It follows that

$$\begin{pmatrix}E_j^{(R+)}\\E_j^{(R-)}\end{pmatrix} = \frac{1}{\det\mathbb{T}^R_j}\begin{pmatrix}\left(\mathbb{T}^R_j\right)_{2,2} & -\left(\mathbb{T}^R_j\right)_{1,2}\\ -\left(\mathbb{T}^R_j\right)_{2,1} & \left(\mathbb{T}^R_j\right)_{1,1}\end{pmatrix}\begin{pmatrix}1\\0\end{pmatrix} \tag{C.22}$$

and hence

$$E_j^{(R+)} = \frac{\left(\mathbb{T}^R_j\right)_{2,2}}{\det\mathbb{T}^R_j},\quad E_j^{(R-)} = -\frac{\left(\mathbb{T}^L_j\right)_{2,1}}{\det\mathbb{T}^R_j}. \tag{C.23}$$

According to (C.10), a left-going field should emerge from slice 1; again, we will assign it a magnitude of unity. Recalling that, in this formalism, the field is propagated from left to right, we deduce:

$$\begin{aligned}\begin{pmatrix}E_j^{(L+)}\\E_j^{(L-)}\end{pmatrix} &= \prod_{i=1}^{j-1} T^L_{i\to i+1}\begin{pmatrix}0\\1\end{pmatrix}\\ &= T^L_{j-1\to i}\cdots T^L_{2\to 3}\cdot T^L_{1\to 2}\begin{pmatrix}0\\1\end{pmatrix}\\ &= \mathbb{T}^L_j\begin{pmatrix}0\\1\end{pmatrix}.\end{aligned} \tag{C.24}$$

The multiplication terminates at $i_0-1$, as this brings the field propagation to a close at the left-hand boundary of cell $i_0$, as required. It follows that

$$E_j^{(L+)} = \left(\mathbb{T}^L_j\right)_{1,2},\quad E_j^{(L-)} = \left(\mathbb{T}^L_j\right)_{2,2}. \tag{C.25}$$

To define the transfer matrix $T^L_{i\to i+1}$, in this case, we note that the field on the left of cell $i$ must be evolved through the medium of cell $i$ before the next set of boundary conditions are applied, and that consequently we must redefine the combined transfer matrix:

$$T^{L}_{i\to i+1} = t_{i\to i+1}\Lambda_i. \tag{C.26}$$

We can formulate different expressions for the field in a cell, which hold up to a multiplicative constant:

$$\begin{aligned} f_l(i,q) &= E_i^{(L+)}\mathrm{e}^{-w_i q} + E_i^{(L-)}\mathrm{e}^{w_i q}, \\ f_r(i,q) &= E_i^{(R+)}\mathrm{e}^{w_i q} + E_i^{(R-)}\mathrm{e}^{-w_i q}, \end{aligned}$$

where $q$ is a displacement of the field from the left-hand-side of the cell. Applying equations (C.23) and (C.25), it follows that

$$f_l(i,q) = \left(\left(\mathbb{T}^L_j\right)_{1,2} \left(\mathbb{T}^L_j\right)_{2,2}\right)\begin{pmatrix}\mathrm{e}^{-w_i q}\\ \mathrm{e}^{w_i q}\end{pmatrix}, \tag{C.27}$$

$$f_r(i,q) = \frac{1}{\det\mathbb{T}^R_j}\left(\left(\mathbb{T}^R_j\right)_{2,2} \left(\mathbb{T}^R_j\right)_{2,1}\right)\begin{pmatrix}\mathrm{e}^{-w_i q}\\ \mathrm{e}^{w_i q}\end{pmatrix}. \tag{C.28}$$

For $g_l$, the field is evolved forward to position $q$. For $g_r$, it must be evolved backwards (from the right-hand-side) of the cell to position $q$. To map these fields to the real axis, we create a selection function:

$$b(x) = \begin{cases} 1 & X_1 \le x < X_2 \\ 2 & X_1 \le x < X_2 \\ \dots & \dots \\ t & X_{t-1} \le x < X_t \end{cases} \tag{C.29}$$

Finally, we introduce the scalar Green function for the field in the system:

$$g_T(x,x_0) = \begin{cases} d_1 g_l(x) & x_1 \le x \le x_0 \\ d_2 g_r(x) & x_0 < x \le x_1 \end{cases} \tag{C.30}$$

where

$$\begin{aligned} g_l(x) &= f_l(b(x), x - X\,[b(x)]), \\ g_r(x) &= f_r(b(x), x - X\,[b(x)]). \end{aligned}$$

The constants $d_1$ and $d_2$ may be found by applying equations (C.4) and (C.8), replacing $c_1$ with $d_1$ and $c_2$ with $d_2$, and solving simultaneously.

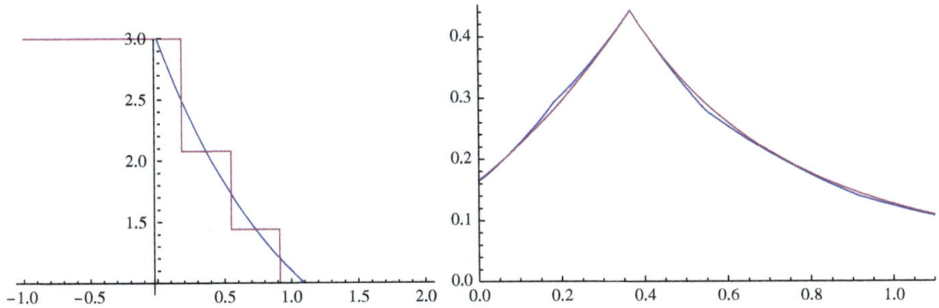


**Fig. C.1** Low resolution comparison. *Left* The refractive index profile. The *blue line* is a plot of the particular profile $n(x)$ used in this example. The purple line is a plot of the values of $n(x)$ in this piecewise approximation. *Right* The *blue line* is a plot of $g_N$, and the *purple line* is a plot of $g_T$, for the case $x_0 = (\text{Log}[3])/3$, $u = 1$, $t = 4$

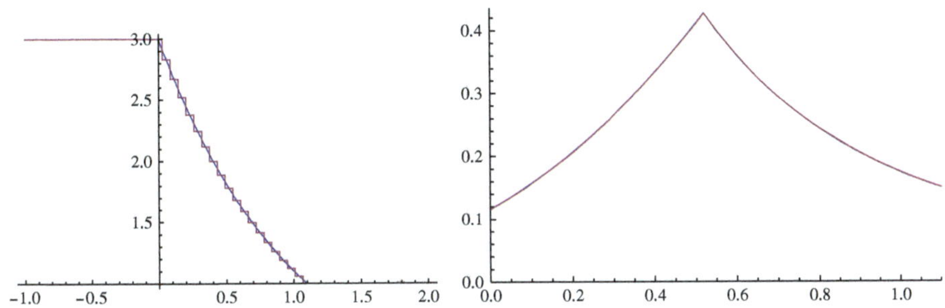


**Fig. C.2** Moderate resolution: $g_N$ and $g_T$ for the case $x_0 = (9\text{Log}[3])/19$, $u = 1$, $t = 20$. The legend is otherwise identical to Fig. C.1

## C.3 Comparison

Let $g_N$ denote the scalar Green function recovered from a numerical solution to the wave equation, as described in Sect. C.1, and let $g_T$ denote the Green function determined using the transfer matrix method in Sect. C.2. Comparative plots demonstrate that the transfer matrix method outlined above recovers the numerical solution of the wave equation, as the number of slices is increased. In Fig. C.1, the number of slices is very small, and $g_T$ is a crude fit for $g_N$. However, using only 20 slices, $g_T$ is already a much better approximation, and is difficult to distinguish from the plot of $g_T$ in Fig. C.2. We also include a second case, involving different wave numbers, in Figs. C.3 and C.4. The accuracy of the approximation $g_T$ continues to improve as the number of slices in the stack is increased.

To recapitulate briefly, we have seen that the wave equation for the case of an inhomogeneous profile can be solved numerically, and that the solution obtained using a numerical solver may be approximated to an arbitrary degree of accuracy

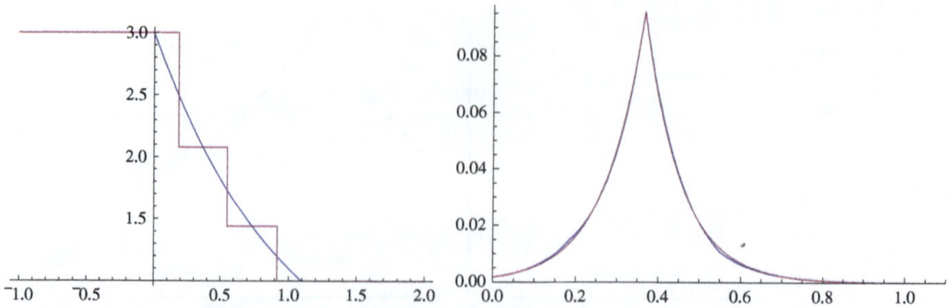


**Fig. C.3** Low resolution: $g_N$ and $g_T$ for the case $x_0 = (\text{Log}[3])/3$, $u = 1$, $t = 4$. The legend is otherwise identical to Fig. C.1

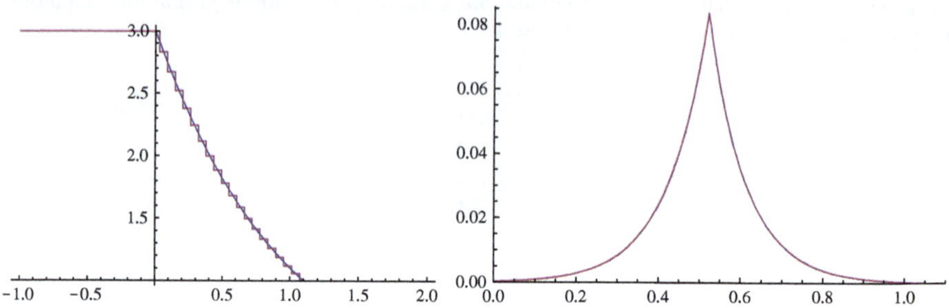


**Fig. C.4** Moderate resolution: $g_N$ and $g_T$ for the case $x_0 = (9\text{Log}[3])/19$, $u = 1$, $t = 20$. The legend is otherwise identical to Fig. C.1

using a piecewise approximation of the profile and determining the relevant transfer matrices.

## Reference

1. U. Leonhardt, *Essential Quantum Optics* (Cambridge University Press, Cambridge, 2010)

# Appendix D
# Möbius Transformations

A Möbius transformation[5] is a mapping in the extended complex plane[6] $\tilde{\mathbb{C}}$ of the form

$$w(z) = \frac{az + b}{cz + d}, \quad w, z, a, b, c, d \in \tilde{\mathbb{C}}. \tag{D.1}$$

## D.1 Theorems on Möbius Transformations

**Theorem D.1** *Möbius transformations are generated (under composition) by*

1. *translations—maps of the form* $z \mapsto z + k$, $k \in \tilde{\mathbb{C}}$.
2. *scalings (or dilations)—*$z \mapsto kz$, $k \in \mathbb{C}\backslash\{0\}$,
3. *inversions—*$z \mapsto 1/z$,

*Proof* Any Möbius transformation can be expressed as a chain of elementary transformations, $w = z_4(z_3(z_2(z_1(z))))$,

$$w = z_4 = z_3 + \frac{a}{c}, \quad z_3 = \frac{bc - ad}{c^2} z_2, \quad z_2 = \frac{1}{z_1}, \quad z_1 = z + \frac{d}{c}. \tag{D.2}$$

$z_1(z)$ and $z_4(z)$ are translations, $z_2(z)$ is an inversion, $z_3(z)$ is a scaling. Clearly, Möbius transformations form a subset of the group of transformations generated by translations, scalings and inversion. □

**Theorem D.2** *Möbius transformations are bijections on* $\tilde{\mathbb{C}}$.

*Proof* Möbius transformations are compositions of translations, scalings and inversions (see Theorem D.1). Since

[5] This section is indebted to the discussion of Möbius transformations in [1].

[6] The extended complex plane is defined as the set $\mathbb{C} \cup \{\infty\}$, denoted $\tilde{\mathbb{C}}$.



W.M.R. Simpson, *Surprises in Theoretical Casimir Physics*, Springer Theses,
DOI 10.1007/978-3-319-09315-4

1. the translations are bijections: $z \mapsto z + k$, $\infty \mapsto \infty$ with inverse $k \mapsto z - k$, $\infty \mapsto \infty$,
2. the scalings are bijections: $z \mapsto kz$, $\infty \mapsto \infty$, with inverse $z \mapsto z/k$, $\infty \mapsto \infty$,
3. the inversions are self-inverting bijections: $z \mapsto 1/z$, $0 \mapsto \infty$, $\infty \mapsto 0$.

A Möbius transformation is also a bijection. □

**Theorem D.3** *Möbius transformations map circles to circles.*

*Proof* Möbius transformations are compositions of translations, scalings and inversions (see Theorem D.1). It is clear that translations and scalings preserve circles. Consider then the inversion $w = 1/z$. A circle (in the domain) of radius $r_0$ centred on the $x$–axis is described by the equation

$$|z - x_0|^2 = r_0^2 \quad \Rightarrow \quad r_0^2 = x_0^2 - x_0(z + z^*) + |z|^2. \tag{D.3}$$

Consider the equation of a circle (in the image),

$$|w - x_0'|^2 = r_0'^2 \quad \Rightarrow \quad |1/z - x_0'|^2 = r_0'^2 \quad \Rightarrow \quad |1 - x_0'z|^2 = r_0'^2|z|^2 \tag{D.4}$$

Thus,

$$\begin{aligned} r_0'^2|z|^2 &= |1 - x_0'z|^2 = 1 + x_0'^2|z|^2 - x_0'(z + z^*) \\ &= 1 - \tfrac{x_0'}{x_0}(x_0{}^2 - r_0{}^2 + |z|^2) + x_0'^2|z|^2 \end{aligned} \tag{D.5}$$

Then, with the parameters

$$x_0' = \frac{x_0}{x_0^2 - r_0^2}, \quad r_0' = \frac{r_0}{|x_0^2 - r_0^2|} \tag{D.6}$$

the equation of the transformed circle is satisfied. For $x_0^2 = r_0^2$ the circles in $z$-space are transformed into straight lines in $w$-space, which we may regard as circles with infinite radius. Conversely, straight lines in $z$-space are transformed into circles in $w$-space, as the same argument may be applied for $z = 1/w$. Thus circles map into circles for real $x_0$, for circles centred at the $x$-axis. We can generalise by rotation for any centre point.

## D.2 Circles on the Sphere in Stereographic Projection

Here, we shall establish that circles are mapped to circles in the stereographic projection. We begin by imagining a circle centred at the North Pole of the sphere. It is obvious that this will project as a circle onto the plane; the stereographic projection is rotationally symmetric. We now imagine rotating the circle on the sphere (that is, allowing its centre to slide from the North Pole to some other point on the sphere's surface). By so doing, we can reach any other circle on the sphere of the same radius.

As it happens, rotation on the sphere corresponds to a particular Möbius transformation (D.1) of the projected points (7.2.6) in the complex plane. The transformation is

$$z' = \frac{az+b}{cz+d} \tag{D.7}$$

where

$$a = \exp^{i\beta}\cos\gamma, \quad b = -\exp^{i\beta}\exp^{i\alpha}\sin\gamma \tag{D.8a}$$

$$c = \exp^{-i\alpha}\sin\gamma, \quad d = \cos\gamma. \tag{D.8b}$$

so that

$$z' = \exp^{i\beta}\frac{z\cos\gamma - \exp^{i\alpha}\sin\gamma}{z\exp^{-i\alpha}\sin\gamma + \cos\gamma}. \tag{D.9}$$

It is a mapping from one set of points in the plane to another. For ease of analysis, we can rewrite this as

$$z' = \exp^{i\alpha+i\beta} z_0'(z\exp^{-i\alpha}), \quad z_0 = \frac{z\cos\gamma - \sin\gamma}{z\sin\gamma + \cos\gamma}, \tag{D.10}$$

where $z_0$ denotes the Möbius transformation (D.9) with $\alpha = \beta = 0$. Clearly, the angle $\alpha$ rotates the points on the $z$-plane, the transformation $z_0'$ is effected, and then followed by another rotation of $-(\alpha+\beta)$. Consider the stereographic projection in inverse (7.2.7),

$$X + iY = \frac{2z}{1+|z|^2}, \quad Z = \frac{|z|^2-1}{|z|^2+1}. \tag{D.11}$$

For rotations on the complex plane $z \to z\exp^{i\theta}$, the spherical coordinate $Z$ obviously remains unchanged, but $X$ and $Y$ are transformed:

$$X + iY = \frac{1}{1+|z|^2}\left(z\exp^{i\theta}\cos\alpha + iz^{i\theta}\sin\alpha\right) \Rightarrow \tag{D.12}$$

$$\begin{pmatrix} X \\ Y \end{pmatrix} = \frac{2}{1+|z|^2}\begin{pmatrix} \cos\theta & -\sin\theta \\ \sin\theta & \cos\theta \end{pmatrix}\begin{pmatrix} x \\ y \end{pmatrix}. \tag{D.13}$$

Thus rotations on the complex plane correspond to rotations of the sphere around the $Z$-axis; the circles on the sphere are preserved. It remains then to consider the effect of $z_0'$, i.e. the transformation (D.9) for $\alpha = \beta = 0$,

$$z \to z' = \frac{z\cos\gamma - \sin\gamma}{z\sin\gamma + \cos\gamma}. \tag{D.14}$$

Considering again the inverse stereographic projection (7.2.7), using

$$X' = \frac{2\text{Re}(z')}{1+|z'|^2}, \quad Y' = -\frac{2\text{Im}(z')}{1+|z'|^2} \quad Z' = \frac{|z'|^2-1}{|z'|^2+1}, \tag{D.15}$$

and having computed

$$\begin{aligned} |z'|^2+1 &= \tfrac{1}{P}\left(|z|^2+1\right), \\ |z'|^2-1 &= \tfrac{1}{P}\left(|z|^2\cos 2\gamma - \cos 2\gamma - 2\text{Re}(z)\sin 2\gamma\right), \\ 2\text{Re}(z') &= \tfrac{1}{P}(|z|^2\sin 2\gamma + 2\cos 2\gamma\text{Re}(z) - \sin 2\gamma), \end{aligned} \tag{D.16}$$

where

$$P = |z|^2\sin^2\gamma + \cos^2\gamma + \text{Re}(z)\sin 2\gamma, \tag{D.17}$$

we find that

$$\begin{aligned} X' &= X\cos 2\gamma + Z\sin 2\gamma, \\ Y' &= Y, \\ Z' &= -X\sin 2\gamma + Z\cos 2\gamma, \end{aligned} \tag{D.18}$$

yielding

$$\begin{pmatrix} X' \\ Z' \end{pmatrix} = \frac{2}{1+|z|^2}\begin{pmatrix} \cos 2\gamma & \sin 2\gamma \\ -\sin 2\gamma & \cos 2\gamma \end{pmatrix}\begin{pmatrix} X \\ Z \end{pmatrix}. \tag{D.19}$$

Clearly, this describes a rotation around the $Y$-axis of the sphere. Taken altogether then, the Möbius transformation (D.9) describes a rotation around the $Z$-axis by the angle $\alpha$, following by a rotation around the $Y$-axis by $\gamma$, ending with a rotation again about the $Z$-axis by $-(\alpha+\beta)$; these correspond to the Euler angles that describe any rotation on the sphere. We are now at liberty to argue that since

1. we can begin with a circle in the complex plane (projected from the circle centred at the North Pole),
2. Möbius transformations map circles to circles, and
3. rotation on the sphere (by means of which we can reproduce all the great circles of the light rays in the virtual space) corresponds to a particular Möbius transformation

we conclude that the light rays in Maxwell's fish-eye follow circles in physical space (as well as virtual space).

## Reference

1. U. Leonhardt, T.G. Philbin, *Geometry and Light: The Science of Invisibility* (Dover, New York, 2010)

MIX
Papier aus verantwortungsvollen Quellen
Paper from responsible sources
FSC® C105338

If you have any concerns about our products,
you can contact us on
**ProductSafety@springernature.com**

In case Publisher is established outside the EU,
the EU authorized representative is:
**Springer Nature Customer Service Center GmbH**
**Europaplatz 3, 69115 Heidelberg, Germany**

Printed by Libri Plureos GmbH
in Hamburg, Germany